Hands-On Palaeontology

Also available from Dunedin:

Terrestrial Conservation Lagerstatten:
Windows into the Evolution of Life on Land (2017)
Edited by Nicholas Fraser and Hans-Dieter Sues
ISBN: 9781780460147

Introducing Palaeontology
Second Edition (2019)
Patrick Wyse Jackson
ISBN: 9781780460833

For further details of these and other Dunedin
Earth and Environmental Sciences titles see
www.dunedinacademicpress.co.uk

HANDS-ON PALAEONTOLOGY
a practical manual

Stephen K. Donovan

DUNEDIN

EDINBURGH ◆ LONDON

Published by
Dunedin Academic Press Ltd
Hudson House
8 Albany Street
Edinburgh EH1 3QB
Scotland

www.dunedinacademicpress.co.uk

ISBNs
9781780460970 (Paperback)
9781780466644 (ePub)
9781780466651 (Kindle)
9781780466668 (PDF)

British Library Cataloguing in Publication data
A catalogue record for this book is available from the British Library

Typeset by Kerrie Moncur Design & Typesetting

Printed and bound in Poland by Hussar Books

*For the Palaeontographical Society, 175 years young in 2022
and an inspiration for all palaeontologists everywhere*

Contents

Getting started

Some theoretical aspects

Working on your collection at home

The wider field: getting involved

Fossils in many fields

Acknowledgements

My first acknowledgement must go to the Government of the Netherlands. I worked in Leiden. Safe in the knowledge that my retirement date was 2nd June, 2021, I wrote the proposal for this book, which was accepted by Dunedin Academic Press (DAP). About two years to complete the book looked ample. And then the Government changed my date of retirement to 2nd October, 2020. Lopping eight months off the time available to write was not really welcomed as such, but it did concentrate the mind wonderfully. Book writing was something that occurred at lunchtimes and weekends, which had to become more focused. Then came the Covid19 pandemic and I was writing at home every day. The last ten or so chapters were written in weeks, not months, thanks to my focused approach. Thanks, too, to the anonymous reviewer of this early incarnation for their constructive comments. Special thanks to Patrick N. Wyse Jackson (Trinity College, Dublin) for reading a later version of the text and spotting too many gremlins, all minor, but enough to make me say 'Oops'!

Many thanks to all the colleagues, reviewers and editors of almost 40 years, many of which I am proud to call friend, who have supported, criticized, corrected and debated my ideas on palaeontology and geology. Anthony Kinahan at DAP and his external assessor both liked my book proposal, and said full steam ahead. I have not been too much of a strain on the children, Hannah and Pelham, during the course of this writing project, mainly because they have been living and studying in Manchester for most of this period. Similarly, my partner, Karen, although providing support and encouragement from across the North Sea, saw me only once or twice a month, at best, and then came Covid19.

Most of the first draft of each chapter was written in one or more café. The Vascobelo Café in Scheltema bookstore in Amsterdam was the place of choice for breakfast and writing on Saturday mornings. I miss those scrambled eggs and coffee now I have retired and moved back to the UK. Lunchtimes in Leiden were shared between Subway (steak and cheese), Wok Your Way (beef teriyaki) and the Hotspot Central (beef Szechwan), which were my favourite places for lunch and writing at lunchtime. McDonald's and, particularly, Subway in Hoofddorp were favourites for scribbling during a light evening meal. All of the above were indispensible, providing a table and coffee (usually more than a coffee) to grease my writing. It is not an exaggeration to say all were essential as I wrote this book.

Stephen K. Donovan
Thursday, 18th March, 2021

Sources

With few notable exceptions, all the illustrations in this book are mine, either new or reproduced from my published papers. Full bibliographic references are provided in the captions of each and every image or diagram that has been previously published. My own illustrations, reproduced herein, appeared in the following research journals and magazines, some sadly no longer with us, which I am grateful to acknowledge for publishing my research papers and more general articles.

Atlantic Geology (Fig. 1.1)

Bulletin of the Geological Society of Norfolk (Fig. 47.2A, B)

Bulletin of the Mizunami Fossil Museum (Figs 8.1B, 10.1, 23.2, 28.1, 32.3, 47.3A, B)

Caribbean Journal of Science (Figs 5.2, 7.1, 25.1C)

Cave & Karst Science (Fig. 13.1)

Contributions to Tertiary & Quaternary Geology (Table 20.1, Figs 20.2, 22.1)

Publication of the Craven & Pendle Geological Society, UK (Fig. 50.4A-C, E)

Deposits (Figs 16.1, 24.2, 25.1D, 30.1, 52.2C, 52.3)

Fossil Forum (Fig. 29.3)

Geological Curator (Fig. 17.1)

Geological Journal (Figs 24.1, 25.1B)

Geological Magazine (Figs 13.2, 23.3, 32.1)

Geology Today (Figs 6.1, 16.2, 45.2–45.7, 52.1)

Ichnos (Figs 14.2, 24.4)

Journal of Paleontology (Fig. 20.1)

Journal of the Geological Society of Jamaica (Figs 29.1, 29.2)

Lethaia (Figs 3.1, 5.3, 9.1, 9.2, 52.6)

Mercian Geologist (Fig. 50.4)

Monographs of the Palaeontographical Society (Fig. 29.4)

Netherlands Journal of Geoscience (Fig. 10.2)

North West Geologist (Figs 26.1, 50.1)

Palaeontology (Fig. 19.1)

Palaeoworld (Fig. 22.2E)

Proceedings of the Biological Society of Washington (Fig. 2.1)

Proceedings of the Geologists' Association (Figs 3.2, 22F, 31.2, 31.3, 31.4, 46.4, 46.5, 51.3, 51.4, 52.2A, B, 52.4)

Proceedings of the Yorkshire Geological Society (Figs 8.1A, 11.1, 11.2, 14.1, 22A–D, 25.1A, 47.2C, D, 48.2, 48.5, 50.4F)

Scripta Geologica (Figs 47.1, 47.4)

Swiss Journal of Palaeontology (Figs 10.3, 47.3C, D)

Tertiary Research (Figs 13.3, 13.4)

Wight Studies: Proceedings of the Isle of Wight Natural History & Archaeological Society (Figs 49.1, 49.3, 49.4)

Permission to reproduce one figure (Fig. 18.1 herein) from the *Proceedings of the Geologists' Association* was kindly granted by the Professor David R. Bridgland, Chair, Geologists' Association Publication Committee.

Foreword

By Professor David A.T. Harper (Durham University)

Life has been evolving on our planet, and as far as we know, nowhere else, for over 3.8 billion years. Our knowledge and understanding of the increasing complexity and diversity of life through time are based on the remains of once-living organisms, fossils. The assiduous collection and careful extraction and preparation of fossils, be they the soft-bodied wonders of the Cambrian, giant dinosaur skulls or the more humble ammonites, brachiopods, crinoids and trilobites, are the fundament of life's history. Yes, there are many good textbooks that explain the science of palaeontology at various levels of detail, but none quite like this one. Stephen Donovan has, based on over 40 years of experience in the field, provided a refreshingly different and lively dimension to this, the most exciting of the earth sciences.

Why is this book so special? Steve, through 52 short chapters (a successful format adopted in his admirable and highly recommended *Writing for Earth Scientists*), has covered virtually everything you need to know to be a practising palaeontologist. Drawing on his own personal experience, that of colleagues, and the literature, the book both educates and entertains. The first cluster of 17 chapters covers some basic and important concepts: fossils should be carefully collected, properly documented, with detailed locality and stratigraphic data, and with common sense. With each fossil there is a context, be it a sedimentary environment, a stratigraphic section or a historical perspective. Moreover, there are occasions where a photograph rather than the specimen itself will serve as a collection. Once you have built up a fossil collection, now is the time to use it and interpret it. A second group of chapters, 18–25, is more focused on the theoretical background to our subject, introducing the reader to various strands of palaeocology (the study of life of the once-living organism and its associated communities), its preservation and the myriad of behavioural patterns or the traces that organisms produced in life. Chapters 26–33 remind us that there is much to be done at home preparing material, managing it and providing adequate storage. Then, the joy of describing and illustrating your new finds. However, you are not alone. The wider field, and how can you get involved are explained in chapters 34–43, culminating in the importance of publishing your material in the right places.

Why is palaeontology so absorbing and exciting? The final chapters, 44–52, through first the vehicle of the field guide, take the reader through eight very different locations, each with its own story to tell. We learn through a series of vignettes, based again on the author's wide experience, some windows on the history of life, whether revealed in massive quarries, on beaches and cliff exposures or simply in building stones or on pavements. The collection of fossils (whether physically or digital) has been fundamental in understanding more recent events such as human evolution (introducing the Piltdown fake) and those in deep time, for example the evolution of ancient life and its environments in the Carboniferous, Cretaceous, Neogene and Quaternary periods in Europe and the Caribbean.

I am certain that you, like me, will thoroughly enjoy *Hands-on-Palaeontology*. The narrative is authoritative, but light, spiced with anecdotes from the Steve's own career. The illustrations are clear, relevant and well chosen, ranging from vital equipment to some superb fossil specimens – all you need to develop your early career or hobby as a *bone fide* palaeontologist. Now you must turn the pages and explore the ancient worlds of planet Earth.

Professor David A.T. Harper
Principal of Van Mildert College,
Durham University.

Introduction

At a time when the DNA of Neanderthal Man has been sequenced and *Jurassic World* hints at possible future breakthroughs, however improbable, fossils have an undoubtedly high profile. Collecting fossils is a popular hobby. There is a wealth of helpful literature and websites which will help you to put an accurate name on your latest find, and to identify the various component parts of the skeleton. But – what next? How can you curate and conserve your specimen so that it retains its physical integrity and scientific value? What features enable you to interpret your specimen as the remains of a living organism? How was it preserved? Where might you find like-minded collectors? And where can further specimens be found? Too many questions, yet here are the answers. *Hands-On Palaeontology* is a comprehensive introduction to these and many other aspects of palaeontology.

There are many books on palaeontology, aimed at amateurs, undergraduates, research students and other aspiring academics. Perhaps commonest amongst these are what might be called guides to identification, from the general (basic texts on fossil variety and morphology) to the specific (field guides to specific groups, localities or horizons). Many of these are readable, comprehensive and provide good advice. Although I am no longer aspiring, I still enjoy buying and reading such volumes, as each and every one contains something that is new and useful to me. Over the years they have helped keep me informed, as well as educating me about new ideas and research tools, and encouraging me to spread my palaeontological wings.

Hands-On Palaeontology is a different sort of book, composed of numerous short chapters on the subject of palaeontology, but with no intention of being a guide to identification or classification. In this way it will differ from most other volumes on palaeontology available at this level; there is more to the subject than just putting a name on a specimen (however important that is, of course). All chapters will be directly relevant to one or more aspects of palaeontology and aimed at the beginner in the broadest sense, both amateur and undergraduate.

Palaeontologists and geologists are encouraged to use the book as much as a reference as a reader, 'dipping in' to the chapters that contain relevant tips, hints and comments to enable them to improve their understanding of their current interest. The book is intended to be informative, readable and, above all, of practical application for all readers. My approach will be similar to Roy Peter Clark, author of *The Glamour of*

Grammar, when he said 'this book invites you to embrace grammar in a special way, not as a set of rules but as a box of tools' (2010, p. 2). Replace grammar with palaeontology in this quotation and you will have a good idea of the shape of the next 52 chapters.

The structure of *Hands-On Palaeontology* is, perhaps, unusual, comprised of short chapters on aspects of palaeontology of relevance to the beginner, from the esoteric (palaeoecology and preservation) to the practical (storing specimens and photography). My aim has been to make the book a comprehensible reference that the nascent palaeontologist will be able to read from cover to cover, but also more akin to an owner's manual for a car than a textbook. For example, if you have a natural mould that you would like to cast, then read the relevant chapter.

My inspiration for writing *Hands-On Palaeontology* comes from my own experience as an amateur in the mid-1970s. I remember that when I was taking my first tentative steps in the field, there were two slender booklets that told you how to be a palaeontologist (Ford, undated; Dimes & Melville, 1979). These were invaluable to me as I made my first stumbling progress as an amateur and then an undergraduate. There is no modern equivalent, at a time when there is so much more that can be said. *Hands-On Palaeontology* is, therefore, a 'how to' book of a sort that is not otherwise available. It is an adjunct to the available literature, which primarily instructs the reader on how to identify fossils and their skeletal elements. *Hands-On Palaeontology* looks in a different direction. I assume that potential readers already have their textbooks and identification guides on the shelf; instead, I aim to spread the hands-on aspects of palaeontology out before the reader.

References

Clark, R.P. (2010) *The Glamour of Grammar*. Little, Brown and Company, New York.

Dimes, F.G. & Melville, R.V. (1979) *Know the Game: Fossil Collecting*. EP Publishing, Wakefield.

Ford, R.L.E. (undated). *Collector's Guide. No. 3 – Fossil Collecting (Palaeontology)*. Watkins & Doncaster, Welling.

GETTING STARTED

CHAPTER 1

HOW TO COLLECT

To me, this is the obvious place to start this book, but I can imagine there are those who consider it such an obvious topic that they assume that Donovan's brain has become addled in his dotage. Yet I remember an undergraduate student who made a collection of fossils over many weeks during a geological mapping project. I have forgotten his name and it is best to similarly leave the university anonymous. I presume that he had thoughts that his fossils could provide important correlation data for his project. Yet he presented his collection to his supervisor in a shoebox.

The specimens were loose and unlabelled. When asked what came from where, he had no answers. This was after two years of a university education in geology. Thus, if asked who needs to read this chapter, my simple answer is anyone who aims to collect fossils systematically and for a purpose. Those of you with a shoebox collection can skip it.

The popular image of a geologist in the field is of a hammering individual, going hell for leather to smash something of interest out of a rock. In truth, a collector soon discovers that hammering well-indurated (= hard) limestones and sandstones is a mug's game. Fossils that are entombed in these rocks are difficult to collect and, if dug out, will commonly be exhumed as a jigsaw puzzle. See, for example, the Silurian crinoid described by Donovan & Pickerill (1995; Fig. 1.1 herein). It came out of the rock in about a dozen pieces, which I numbered with a fine, black indelible pen and located on a sketch to show what came from where. This specimen was so difficult to hammer and chisel out that I remember having to break off regularly to rest my right (hammer) and left (chisel) hands from jarring.

Back home, I reassembled these pieces with white woodworking glue, which is soluble in water, in a sandbox. (A sandbox is, quite literally, a box of sand used when gluing fossils back together; it permits them to be oriented any way that will favour the action of the glue.) As you'll see from Figure 1.1, this is certainly not the prettiest fossil crinoid ever collected, but a rare species in the Arisaig section of eastern Canada, nonetheless.

It is also possible to go collecting where a hammer is not needed. This might be by beachcombing for reworked fossils (Chapter 47), collecting from sites where the specimens simply drop out of poorly lithified rocks (Chapter 50) or even just using a camera (Chapters 16, 46). Sitting on a scree slope and turning clasts (= fragments of rock) can yield good material. Certainly, specimens from scree, beachcombing or shales may be well preserved

1

Figure 1.1 Monobathrid camerate crinoid gen. et sp. indet., Moydart Formation, Arisaig Group, Upper Silurian (Ludlow), coastal Nova Scotia, eastern Canada (after Donovan & Pickerill, 1995, fig. 3.1). Scale bar represents 10 mm.

and, significantly, will likely be exposed more or less 'in the round'. In contrast, having hammered your fossil out of, say, a lithified limestone, how do you completely expose it? A calcium carbonate (calcite) shell in a calcium carbonate rock (limestone) has no chemical contrast with its substrate; it cannot be freed by acids or other solvents. You will need to invest many hours to release it mechanically, perhaps starting with a hammer and chisel, but certainly ending with a pin in a pin vice. If you are lucky, your shells may be silicified in limestone (Chapter 51) and can be freed with dilute hydrochloric acid, but silicification itself may alter the appearance of a fossil.

So, my philosophy is that you might collect any fossil that you want, but you will need to cut your cloth accordingly. With digital photography providing a cheap and easy way of accumulating multiple images, a wealth of information can be obtained for little cost and no additional bulk; that is, you do not need to collect absolutely every fossil that you see. When I was a student, photography was an expensive adjunct to fieldwork, but times and technologies have changed. This means that one of my interests, 'collecting' fossils from building stones by photography, has become a viable research project. Much of my work on building stones during the past ten years was cheap, but would have been an expensive hobby in, say, the 1980s.

Whether collecting images or fossils or both, you need to know where your specimens came from within a section. I talk about the field notebook in Chapter 4, but you should use it to record relevant details of all of your collecting. Write down where you collected, with grid reference and/or GPS reading, and give each locality a unique number. Write this number in your notebook and also on your sample bags. Each locality needs a number; each bag needs the right number and, preferably, a short descriptor, such as 'Disused quarry above McGregor's Farmhouse, near Snailstrand, county of Clagshire, grey limestones' or whatever. I make similar recommendations in Chapter 4, but if I repeat it enough times someone might just listen!

Collecting bags can be either plastic or linen. Linen bags are preferred by micropalae-ontologists. They collect bulk samples of poorly lithified sedimentary rocks for processing in the laboratory; their fossils are, after all, microscopic and need to be gleaned from big samples under a microscope. Linen bags have the advantage of being porous, so samples can dry, and they can be written on directly with a marker pen. I prefer sealable plastic bags of the sort that can be bought cheaply in supermarkets almost anywhere. Ideally, such bags have a surface that enables you to write directly on them with a ballpoint pen, but, if not, just make whatever notes are essential on a slip of paper and drop it in. My colleague, Dr Fiona Fearnhead, plans ahead; if she knows which locality or localities she will visit in the course of a fieldtrip, she prints off labels with essential data in advance. Now, that is organized!

Use whatever you have to wrap specimens and protect them. Newspaper works well, as do kitchen roll, tissues and toilet paper. The modern newspaper is padded by 'supplements' of diverse sorts, which I rarely read and recycle as wrapping paper. If I am staying in a hotel, I will probably requisition a toilet roll or box of tissues for the field. Always be sure to have some sort of dry wrapping paper with you – multiple specimens from one site, all jostling in the same bag, will grind against each other if they are not protected.

Then you have to get home. Obviously, the smaller the specimens, the less you will have to carry, but a rich site will soon fill a backpack. Pleistocene mammal bones, large Cretaceous and Cenozoic molluscs like oysters and rudists, and mid-Palaeozoic colonial corals are locally common, but bulky, so hopefully your site is near the car park or the railway station. Essentially, if you anticipate collecting the big, then be prepared. Big specimens also fill your collection space – a collection of *Iguanodon* limb bones can be 'big' without including many specimens!

References

Donovan, S.K. & Pickerill, R.K. (1995) A camerate crinoid from the Upper Silurian (Ludlow) Moydart Formation of Nova Scotia, Canada. *Atlantic Geology*, **31**: 81–86.

CHAPTER 2

WHERE TO COLLECT

To start with two 'don'ts' – do not trespass, and do not risk life and limb. The simple answer to 'where' is that you find fossils in fossiliferous rocks. The hard answer is more interesting. Fossiliferous rocks are commonly, almost invariably sedimentary rocks (see Chapters 7 and 8), but there are interesting exceptions. Sedimentary rocks that are involved in mountain-building may become metamorphosed (= metamorphic rocks), such as in North Wales where slates preserve, for example, rare trilobites. A slate is a metamorphosed mudrock; higher grades of metamorphism will likely destroy organic remains, but there are exceptions (see, for example, Laborda-López *et al.*, 2015). Volcanic rocks, particularly ash falls, may act like sand in the marine realm. The Middle Miocene Grand Bay Formation of Carriacou in the Lesser Antilles is highly fossiliferous, but includes fresh mafic mineral grains (Jackson *et al.*, 2008). These tuffaceous sandstones came off the volcano, into the sea and down into deep water.

So, you are searching for an area of fossiliferous sedimentary rocks with accessible exposures. Preferably, your site should be richly fossiliferous – the Old Red Sandstone (Devonian) and New Red Sandstone (Permo-Triassic) are both locally fossiliferous, but both are continental deposits ('red beds', as their names imply). Unless you have a fossiliferous horizon precisely located in a red bed succession, you are unlikely to find any fossils at all.

Focus on your field area and start to hoard information. All sorts of data are likely to be freely available on the Web. The relevant geological map is a must, showing the outcrop area of the sedimentary beds in which you are interested. The outcrop is the surface distribution of a rock unit, largely hidden by soil, fields, towns, forests, roads and the like, but still there just beneath the surface cover. You are looking for surface exposures within this outcrop area, such as the coast, quarries and cuttings, assuming you can get necessary permissions. Field guides can be useful for finding exposures – see, particularly, the long series of *Geologists' Association Guides*, some of which have been through several editions. Best of all, perhaps your local field society is planning a field meeting in your area; you are likely to learn much if you tag along.

All this information is good, but the important thing to do is seek permission to collect from the landowner. Let them know exactly what you want to do, when and how. It is their property – do not expect to always be welcomed. If the answer is no,

Figure 2.1 The heart urchin *Meoma ventricosa* (Lamarck) from the Pleistocene of Barbados, preserved as an internal mould (after Donovan & Jones, 1994, fig. 2). The specimen came from a large boulder lying on the floor of a quarry and is the only fossil specimen of this species known from the island. University of Alberta specimen 9496. **A**. Apical view. **B**. Oral view. Scale bar represents 40 mm.

then that is the answer – just walk away. Move on to the next site. Quarry owners and others will expect you to be responsible, with protective helmet and high-visibility vest. In a working quarry, there will be areas out of bounds to you because they are being worked or are otherwise dangerous. Faces should be avoided because they are probably unstable, but spoil piles are probably more exciting, even if the stratigraphic context is lost (Fig. 2.1).

I enjoy beachcombing. Like quarry faces, I avoid collecting from or even near vertical cliff faces. There may be so much to find on a beach that ignoring the cliffs will be little strain (see, for example, Chapters 14, 47 and 48). Obviously, fossils in clasts on the beach may vary from the fresh to the highly corraded (that is, they have deteriorated by both corrosion and abrasion). I am also fascinated by the interactions of modern with ancient invertebrates, such as encrusting organisms and borings on Upper Cretaceous echinoids on the north Norfolk coast (Donovan & Lewis, 2011; Fig. 47.3C, D). Sites where mud-rocks are exposed in cliffs will likely have some of their fossil content concentrated on the beach, such as the Gault Clay Formation at Folkestone in Kent.

Road cuttings may be fruitful, but take good care of the traffic and make sure that motorists can see you. The best cuttings are those fringed by a grass verge, removing you from the traffic. The Red Hills Road Cave in Jamaica opened straight onto a winding main road and Jamaican drivers are always in a hurry. We would protect ourselves with traffic cones or by parking the vehicle just before the cave. Farley Cutting near Much Wenlock in Shropshire, on the road to Buildwas, has a verge, and the cutting slopes at about 45° on which I commonly lie to bring my short-sighted eyes close to the surface.

Figure 2.2 Two favourite hammers and two favourite chisels, with scales. Note the narrow blades of the chisels, my personal preference. Hammer your chisels, or just use one end or other of the hammer, but never hammer a hammer with a hammer. They are liable to chip; one colleague lost an eye doing this.

The only problem with this is that about one motorist every 15 minutes will stop to make sure that I am OK, bless them.

In some places fossils may just 'come up' in the soil. For example, near Grassington in Yorkshire is a site that yields Mississippian blastoids (extinct, crinoid-like primitive echinoderms), where the thecae occur loose beneath the thin soil. Peel the turf back and there they are, waiting to be collected. Be warned that a blastoid theca is a similar size and shape to a rabbit turd, but not so squishy!

From my comments so far, you will have gathered that I am not an overly active hammerer. I do use a hammer when necessary or, preferably, a hammer and a narrow chisel (Fig. 2.2). I commonly use these to dig a narrow 'trench' around a specimen of interest in a massive rock. With perseverance, this will leave the specimen on a raised plinth. Now use the chisel to dig under the specimen, hammering it around and around, until the specimen pops off on a protective piece of underlying rock. If it breaks into two or more fragments, lose no sleep: I recommend ordinary white woodworking adhesive to fix most damaged fossils.

References

Donovan, S.K. & Jones, B. (1994) Pleistocene echinoids from Bermuda and Barbados. *Proceedings of the Biological Society of Washington,* **107**: 109–113.

Donovan, S.K. & Lewis, D.N. (2011) Strange taphonomy: Late Cretaceous *Echinocorys* (Echinoidea) as a hard substrate in a modern shallow marine environment. *Swiss Journal of Palaeontology,* **130**: 43–51.

Jackson, T.A., Scott, P.W., Donovan, S.K., Pickerill, R.K., Portell, R.W. & Harper, D.A.T. (2008) The volcaniclastic turbidites of the Grand Bay Formation, Carriacou, Grenadines, Lesser Antilles. *Caribbean Journal of Science,* **44**: 116–124.

Laborda-López, C., Aguirre, J. & Donovan, S.K. (2015) Surviving metamorphism: taphonomy of fossil assemblages in marble and calc-silicate schist. *Palaios,* **30**: 668–679.

CHAPTER 3

WHAT TO COLLECT

The answer to this question, assuming you are a beginner, is easy – everything that comes your way. The late Professor H.H. Read of Imperial College, University of London, said that the best geologist is the one who has seen the most rocks. I would further suggest that the best palaeontologist is the one who has collected and studied the most fossils. In collecting a variety of fossils from every site that you visit, of different ages and representing different ancient environmental settings, you will be educating yourself in many key aspects of the science.

But, once collected, if your specimens are put in a shoebox in the shed or under your bed, you will learn very little; ignorance is not the route to being the best palaeontologist that you can be. Getting to know your own collection is an excellent way to develop your own insights. I suggest that this is a three-stage process.

1. Make observations of specimens in the field and record them in your field notebook.
2. Identify and make observations of specimens in the 'laboratory', which may be a grand name for the kitchen table.
3. Read everything of relevance, both geology and palaeontology.

Observing specimens in the field

In the field, you should ask of every specimen, did it live and die here or was it transported from elsewhere? Some organisms have very obviously been preserved where they lived, such as the colonial corals in a fossil reef. Others need more thought and observation, such as fossil acorn barnacles (balanids) encrusting a shell. The barnacles are obviously *in situ*, but where did the shell come from (Fig. 3.1)? Ammonites, fishes and planktonic foraminifera (microfossils) must have settled out of the water column, with some exceptions, and were thus transported.

Consider, as an example, the specimen in Figure 3.2. The heart urchin *Eupatagus* sp. is encrusted by a pair of cementing brachiopods. The brachiopods attached to a dead urchin test are thus autochthonous (see below) with respect to this substrate. In life, the urchin would have been well protected by a dense array of spines and microscopic, jaw-

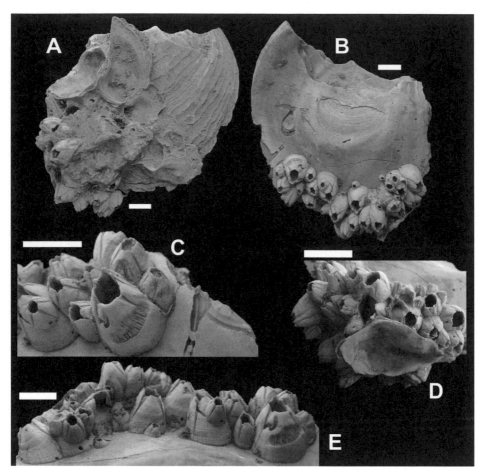

Figure 3.1 Broken oyster valve, *Ostrea edulis* Linnaeus (Naturalis Biodiversity Center, Leiden, specimen RGM 791 569), encrusted by oysters and *Balanus* sp. cf. *B. perforatus* Bruguière, and bored by annelids (after Donovan & Novak, 2015, fig. 1). **A, B.** Outer (**A**) and inner surfaces (**B**) of broken valve showing encrustation by oysters and balanids externally, and balanids internally. The boring towards 10–11 o'clock from the left-hand balanids in (**B**) is *Caulostrepsis taeniola* Clarke. **C.** Detail of balanids and *C. taeniola* on inner surface; note small round holes, *Oichnus simplex* Bromley, near the bases of some balanids. **D.** Balanids at commissure on inner surface and encrusted, in turn, by an oyster. **E.** Balanids encrusting inner surface; note common *O. simplex* low on the shells. Scale bars represent 10 mm. Specimens not coated with ammonium chloride (see Chapter 28).

like structures called pedicellariae that pick off settling larvae. The brachiopods encrusted a bald rather than a spiny test – the echinoid was already dead. The echinoid itself must have been exhumed, at least (= allochthonous; see below), if not further transported by bottom currents, because in life it would have been infaunal and thus removed from the benthic sea floor where the brachiopods lived.

I have a personal fascination with fossil 'sea floors' (such as Donovan, 2014), that is, bedding planes with a rich mixture of fossils. Every specimen poses questions. For

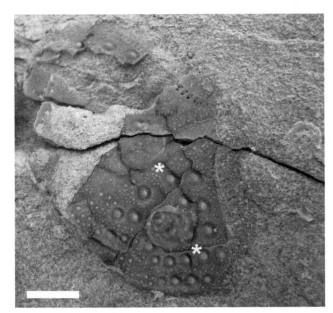

Figure 3.2 Field photograph from Half Moon Bay, Antigua, of University of Florida UF 277885, the apical surface of the spatangoid echinoid *Eupatagus* sp. encrusted by the thecadine brachiopod, thecideid sp. indet. (after Donovan *et al.*, 2017, fig. 3). Attached brachiopod valves are indicated thus (*). Scale bar represents 10 mm.

example, is it *in situ* (autochthonous), locally derived (parautochthonous) or derived by transport from elsewhere (allochthonous)? Most crinoids lived (and live) with the column elevated, but at least part of the distal stem may have been recumbent on the sea floor. So, are the pluricolumnals (= fragments of crinoid column) derived from upright or recumbent crinoids? One source of relevant information comes from the organisms attached to the pluricolumnals. If, for example, a colonial coral encircles the pluricolumnal through 360º, it is probable that the column was upright in life. But an upright crinoid can also be encrusted on a seafloor after death, but not through 360º unless it was rolled around. Consider how many specimens there are in, for example, Figure 50.4D – there are possibilities, probabilities and impossibilities to be determined for each one.

Observing specimens at home

The home palaeontological laboratory (as I said before, to give your kitchen table a new name) can be as simple or as complicated as you make it. The basic equipment includes adequate light (preferably daylight) and your eyes (I talk about storage of specimens in Chapter 26). Eyes can be improved with a hand lens, what a colleague of mine calls a see-bigger-scope. In practice, palaeontologists are accumulators of hand lenses – I have at least half a dozen – and they are, of course, an essential part of your field kit, too. You want to see things bigger to pick out fine details, so a binocular microscope would be of use. Do not break the bank, but if you see a binocular microscope of the type that allows you to examine whole specimens, and it is on sale/a present from a friend who no longer uses it/in a dumpster behind a university or museum, then make it yours.

References

Donovan, S.K. (2014) Palaeoecology and taphonomy of a fossil 'sea floor', in the Carboniferous Limestone of northern England. *Mercian Geologist*, **18**: 171–174.

Donovan, S.K., Harper, D.A.T., Portell, R.W. & Toomey, J.K. (2017) Echinoids as hard substrates: varied examples from the Oligocene of Antigua, Lesser Antilles. *Proceedings of the Geologists' Association*, **128**: 326–331.

Donovan, S.K. & Novak, V. (2015) Site selectivity of predatory borings in Late Pliocene balanid barnacles from south-east Spain. *Lethaia*, **48**: 1–3.

CHAPTER 4

THE FIELD NOTEBOOK

There are many essential items of field equipment (Tucker, 2011), but the most important is your notebook. I have a slip file at home containing old field notebooks that go back almost 40 years. I still refer to them. If you do not write it down, it will drop out of your memory; that is, for you, it will cease to exist. Several years ago I was in the field in the Caribbean with a group of co-researchers. One was measuring a section and collecting specimens from each bed. Another colleague, who enjoyed an argument, said that the section measurer was too slavish in his note-taking and should spend more time committing observations to memory. A few years later, the section-measurer wrote a first draft of a joint research paper, to which the arguer could make no worthwhile contribution as he had nothing in his field notebook and had forgotten the section.

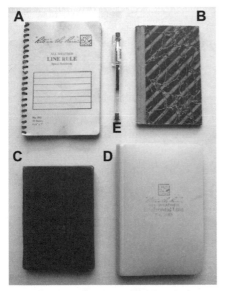

I admit to being old-fashioned in liking to write in a notebook, on paper, in the field. The modern preference for computers, ipads and tablets has not changed me, although I do understand that they have advantages. Yet a paper notebook is lighter than a laptop and never has a flat battery. If there is any intention to publish your findings, do not just upload your field notes. These will always need rewriting before publication. Whatever, although the following is written from the point of view of a paper notebook, my comments should nonetheless also be of use to a computer user in the field.

Figure 4.1 Four of my field notebooks (A–D), going back to the 1980s, and a black pen of the sort with which I enjoy writing (E). The two yellow notebooks (A, D) are Rite in the Rain®; apart from being waterproof, bright yellow makes them more difficult to misplace in the field. Notebooks (B, C) have the advantage of slightly smaller in size and being a sort that is widely available in stationery stores. The pen (E), a Pilot G-Tec-C4, has a fine nib and is 135 mm in length.

Field notebooks and portable computers come in many shapes and sizes, depending upon the taste of the note taker (Fig. 4.1). I abhor A4 and A5 notebooks, which I find too cumbersome, but there are those who swear by them. I favour something smaller than A5 that is handy to slip into a pocket. Stiff covers are good, giving you a surface to push against when writing. I also like a notebook that is flexible, allowing it to be doubled-over, producing a hand-sized writing surface. A ring-bound notebook is best for this. Ideally, at least the covers should be waterproof. The American company, J.L. Darling, make Rite in the Rain® waterproof notebooks, a favourite of mine (Fig. 4.1A, D), their product having plastic covers, waterproof paper and, if you want it, ring bindings.

What you use to write is also a personal preference. I always endeavour to keep my notebook dry, so there is little chance that ink will run. So, a fine-nibbed pen of some description is good for me (Fig. 4.1E). I dislike ballpoint pens, and a field notebook is not the right place for me to use a broad nib. Pencils are excellent, but make them hard, please. That is, not a HB, but a 2H or harder still; somewhere at home I have a 10H. I have met particularly diligent fieldworkers who write in pencil in the field and then ink over their notes, for permanence, later that evening.

So, once you have chosen your field notebook and a preferred writing implement, what do you write? Pretty well anything and everything of relevance should be noted (Fig. 4.2). The date is a good start and, as an aide-mémoire, a terse note or two about the weather and how you got there. 'Oh, yes, I remember that trip, brilliant sunshine, but the trains were disrupted because...'

I recommend that every locality has a unique number. My method is to have the date followed by a number. Today is 10th July 2019, so if I was in the field right now, my first locality would be 10/7/2019/1 and so on. Use these numbers on your sample bags, too, for ease of cross referencing. A simple alternative is just to give every site a sequential number. When I worked at the University of the West Indies, Professor Edward Robinson was up to locality 2500 or thereabouts – and that was over 20 years ago!

You will develop your own methods of note-taking. The trick is to write concisely, but not so tersely that you cannot decipher your own notes six months later. There are some standard abbreviations that we all use, such as the following for sedimentary rock types: lst = limestone; mst = mudstone; sst = sandstone; and cong = conglomerate. Coal and shale are short words and need no abbreviating.

Maps and diagrams are always useful after the event. These are not destined for an art gallery; they are to remind you how it was on the day. They are for your use; nobody else will likely ever see them. Sketch maps will be just that; never forget to include a scale and a north arrow. Diagrams cover a multitude of possibilities, from a specimen to a mountain or a glaciated valley. Again, do not forget the scale – ever! Also note the orientation, such as 'looking north'. Measured sections (Figs 4.2, 4.3) are discussed in detail in the next chapter.

Digital photography is cheap and easy, but it is not a replacement for an accurate sketch, although it may be a worthy adjunct. Sketching, like riding a bicycle, improves with practice. So, before you say that you cannot draw, I will say that the more diagrams

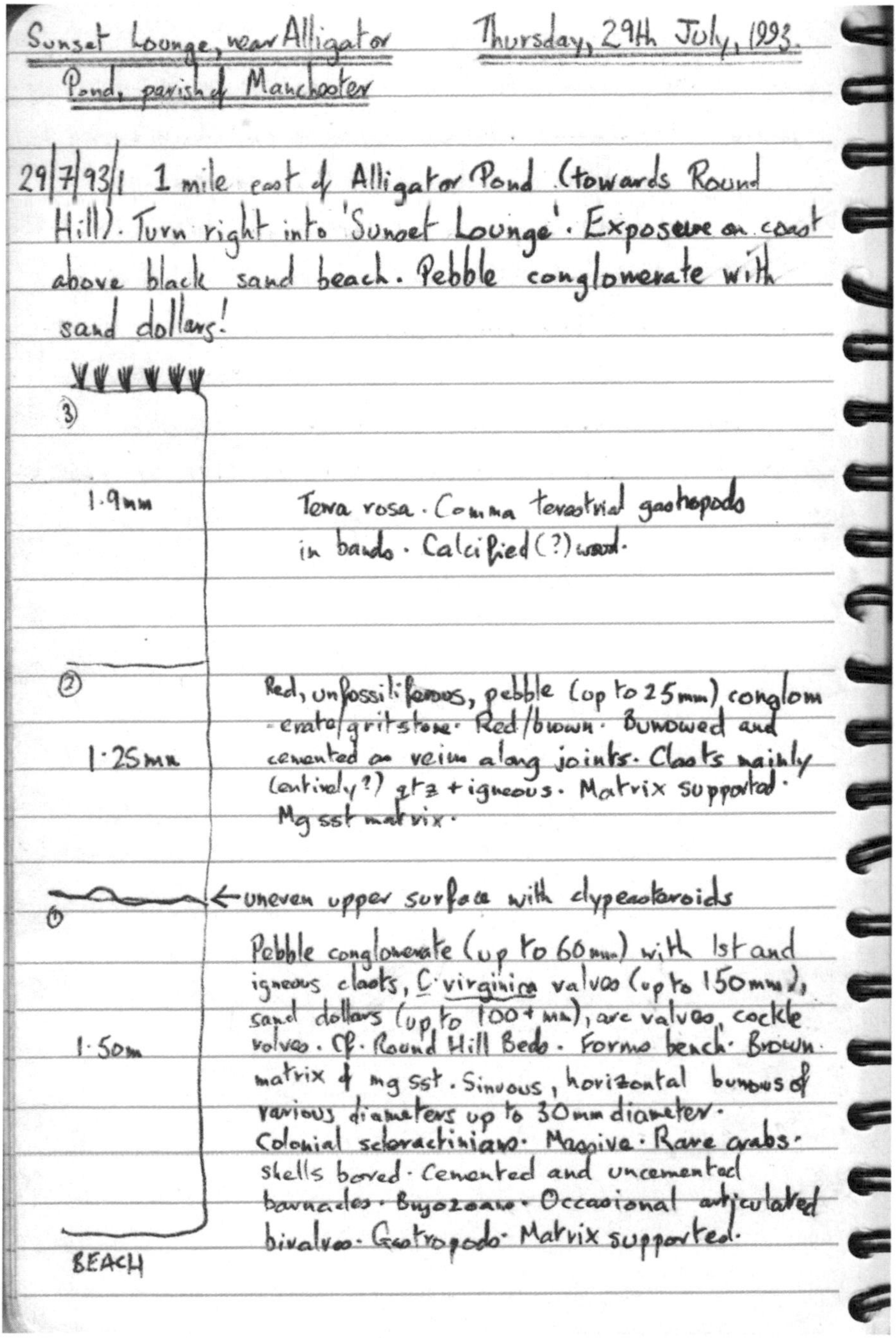

Sunset Lounge, near Alligator   Thursday, 29th July, 1993.
Pond, parish of Manchester

29/7/93/1   1 mile east of Alligator Pond (towards Round
Hill). Turn right into 'Sunset Lounge'. Exposure on coast
above black sand beach. Pebble conglomerate with
sand dollars!

ⅴⅴⅴⅴⅴⅴ

③

1.9mm   Terra rosa. Comma terrestrial gastropods
             in bands. Calcified (?) wood.

②   Red, unfossiliferous, pebble (up to 25mm) conglom
             -erate/gritstone. Red/brown. Burrowed and
1.25mm   cemented or veins along joints. Clasts mainly
             (entirely?) qtz + igneous. Matrix supported.
             Mg sst matrix.

①   ←uneven upper surface with clypeasteroids
             Pebble conglomerate (up to 60mm) with 1st and
             igneous clasts, C. virginica valves (up to 150mm),
             sand dollars (up to 100+mm), arc valves, cockle
1.50m   valves. Cf. Round Hill Beds. Forms beach. Brown.
             matrix of mg sst. Sinuous, horizontal burrows of
             various diameters up to 30mm diameter.
             Colonial scleractinians. Massive. Rare crabs.
             shells bored. Cemented and uncemented
             barnacles. Bryozoans. Occasional articulated
             bivalves. Gastropods. Matrix supported.

BEACH

Figure 4.2 A typical page from a notebook (Fig. 4.1A), showing a measured section in Jamaica that was eventually published (Donovan *et al.*, 1994, fig. 3). This was the culmination of a project to rediscover the type locality of the Neogene sand dollar *Encope homala* Arnold & Clark. The published locality data stated simply parish of Manchester in Jamaica; this is equivalent to, say, the county of Sussex in the UK.

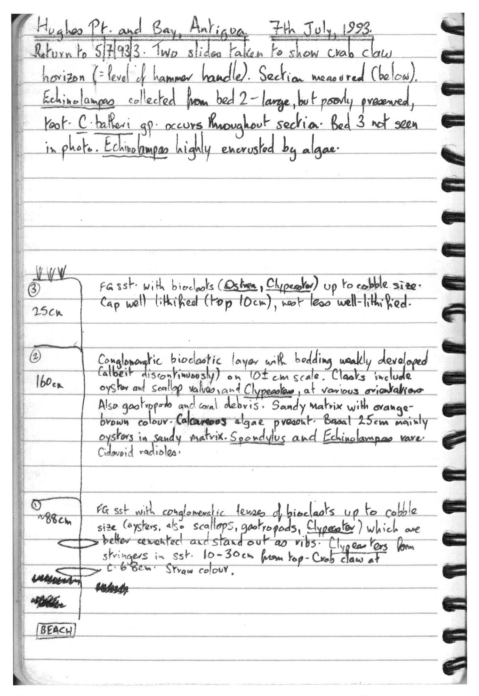

Hughes Pt. and Bay, Antigua   7th July, 1993.
Return to 5/7/93. Two slides taken to show crab claw
horizon (= level of hammer handle). Section measured (below).
Echinolampas collected from bed 2 - large, but poorly preserved,
root. C. batheri gp. occurs throughout section. Bed 3 not seen
in photo. Echinolampas highly encrusted by algae.

③ 25cm
FG sst. with bioclasts (Ostrea, Clypeaster) up to cobble size.
Cap well lithified (top 10cm), rest less well-lithified.

② 160cm
Conglomeratic bioclastic layer with bedding weakly developed
(albeit discontinuously) on 10± cm scale. Clasts include
oyster and scallop valves, and Clypeaster, at various orientations.
Also gastropods and coral debris. Sandy matrix with orange-
brown colour. Calcareous algae present. Basal 25cm mainly
oysters in sandy matrix. Spondylus and Echinolampas rare.
Cidaroid radioles.

① ~88cm
FG sst with conglomeratic lenses of bioclasts up to cobble
size (oysters, also scallops, gastropods, Clypeaster) which are
better cemented and stand out as ribs. Clypeasters form
stringers in sst. 10-30cm from top. Crab claw at
c. 68cm. Straw colour.

BEACH

Figure 4.3 Another typical page from a notebook, showing the date, locality
(in Antigua), some terse notes and a sketch measured section. This was an
exciting day in the field, as I discovered the first fossil crab from Antigua. This
was published soon after (Collins & Donovan, 1995; see Fig. 5.2 herein).

you put in your notebook, the better they will become with time. The sketches in your field notebook are for you, your memory, your observations and your ideas. Paintings for the National Gallery are something else and are, to the best of my knowledge, never straight out of a field notebook (although perhaps they started on their pages). The focus of your sketches is what you see and want to record.

References

Collins, J.S.H. & Donovan, S.K. (1995) A new species of *Necronectes* (Decapoda) from the Upper Oligocene of Antigua. *Caribbean Journal of Science*, **31**: 122–127.

Donovan, S.K., Dixon, H.L., Littlewood, D.T.J., Milsom, C.V. & Norman, Y.J.C. (1994) The clypeasteroid echinoid *Encope homala* Arnold and Clark, 1934, in the Cenozoic of Jamaica. *Caribbean Journal of Science*, **30**: 171–180.

Tucker, M.E. (2011) *Sedimentary Rocks in the Field: A Practical Guide*. Fourth edition. Wiley-Blackwell, Chichester.

CHAPTER 5

MEASURING SECTIONS (AND WHY)

I have taken a relaxed approach to our collecting so far. Make sure your specimens are bagged with adequate locality data. They should be treated equally whether they were collected *in situ* or were float, loose on the floor of a quarry or on the beach. In this chapter I shall focus on the *in situ* specimen and its added value. Such a fossil provides context that is lost in a float specimen. By context I am referring to both time and space. Time refers to the height of your section; specimens lower in the section are older and those from higher are similarly younger assuming beds have not been overturned (see Chapter 6). With such information the collector can start to build up a feel for faunal succession – what came first, what came after. Space refers to what other species are preserved at the same bedding horizon as your specimen. Accurate collection of specimens from one bed or bedding plane will permit you to start forming ideas of the associations of organisms to be found at one brief instant of time.

An obvious problem is how do you record exactly where in a section a specimen was collected? You want to make a record that is not only clear to you if you return to collect more after, say, five years, but will also be clear if you direct a colleague to the same site or bed. Just where the locality is must be recorded as accurately as possible in your field notebook (see Chapter 4). You might record the horizon from which your specimen was collected by photography. Take a close-up photograph of the specimen *in situ* and with a scale of some sort, either a ruler (in cm and mm) or an object of known diameter (such as a coin). Mark this position with a big, obvious object such as your hammer, step back (be careful) and photograph the whole section. If you make the head of the hammer indicate the specimen's position, you will be able to relocate the horizon with ease.

This is the simplest situation, but what of the section from which you collect many specimens? Will a series of similar photographs be usable or just confusing? And how might you combine all of these observations into one coherent whole? The answer is to add one piece of equipment to your field kit – a tape measure – and use it to construct a diagram in your field notebook – a measured section (Figs 4.2, 4.3).

The tape measure is a personal taste, as ever, but I prefer a 5 m, self-winding steel tape that fits easily into the hand and pocket (Fig. 5.1). These are portable, light, tough and

Figure 5.1 My latest tape measure. Fits in the pocket, fits in the hand, will last for years and is essential if you are going to draw measured sections or, indeed, make any other measurements in the field.

Figure 5.2 Measured section of the locality that yielded *Necronectes summus* Collins & Donovan, section exposed near Hughes Point, Nonsuch Bay, east Antigua (after Collins & Donovan, 1995, fig. 2; compare with Fig. 4.3). Beds 1 to 3 were explained in the original paper. The arrow indicates the crab horizon. Key: F, M, C = fine-, medium- and coarse-grained sandstones; P = pebble conglomerate; C = cobble conglomerate.

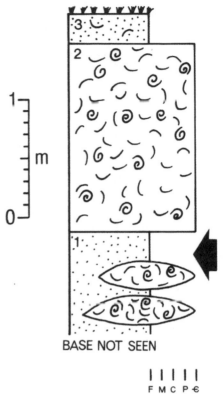

suitable for most jobs in the field. I have known people use the sort of cloth tapes used by tailors or dressmakers; paper tape measures are given away freely in Ikea stores. For some jobs a long measure of tens of metres may be preferred.

A measured section is a graphical representation of a succession of layered rocks (commonly sedimentary, possibly including some volcanic rocks). If you have a sequence of beds, you can measure the thicknesses of each one and construct a representation of each in order. Figure 4.3, from my field notebook of 1993, shows my field sketch section of a short sequence of the Upper Oligocene Antigua Formation of Antigua, Lesser Antilles. The purpose of this section was to show the precise occurrence of what was the first decapod crustacean (crab) found on the island, a large claw, which was later described as *Necronectes summus* Collins & Donovan. The published section (Fig. 5.2) was based on that in the field notebook.

I did not return to the fossil-rich rocks of Antigua until 2013. My colleagues and I found many fossil invertebrates, including echinoids and more (and more complete) crabs, but also other more exotic fossils (at least for the Oligocene of the Antilles), such as crinoids, brachiopods and trace fossils. One of the most productive sections was at Half Moon Bay, in the south-east part of the island. The association of crinoids, brachiopods and thin-walled sponges in rocks of this age is an indication that they were deposited in

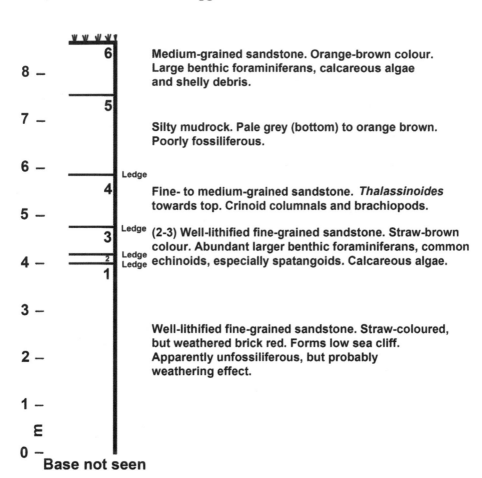

Figure 5.3 Measured section of the north-east point of Half Moon Bay, south-east Antigua, in the Antigua Formation (Upper Oligocene) (after Donovan *et al.*, 2015, fig. 3). Note the section is entirely in limestones; terms such as sandstone and mudrock refer only to grain size. Crinoid columnals and a brachiopod were collected from bed 4; both groups are rare higher in the section.

deep water. The stratigraphic distribution of several fossil groups was summarized in a measured section (Fig. 5.3).

The important points to realize are that measured sections can be drawn in a variety of styles (compare Figs 5.2 and 5.3); different styles may emphasize different physical and lithological features of the section; the value of a measured section is increased by detailed annotations; and all of these are determined by you to illustrate those features that you consider to be of importance. The measured section will enable those who come after you to precisely identify the beds from which you collected. For as long as a rock face remains exposed and accessible, your measured section, especially if published, may remain a standard reference to the occurrence of its palaeontological treasures.

References

Collins, J.S.H. & Donovan, S.K. (1995) A new species of *Necronectes* (Decapoda) from the Upper Oligocene of Antigua. *Caribbean Journal of Science*, **31**: 122–127.

Donovan, S.K., Harper, D.A.T. & Portell, R.W. (2015) In deep water: a crinoid–brachiopod association in the Late Oligocene of Antigua, West Indies. *Lethaia*, **48**: 291–298.

CHAPTER 6

THE LAW OF SUPERPOSITION

It is time for some light theory of stratigraphy—nothing too heavy, but essential none-theless. A more complete title for this chapter might be 'Some technical stuff'. We will start with the Law of Superposition and then become further diverted.

The Law of Superposition

Much of what I said in the first five chapters was underpinned by the Law of Superposition; it is so obvious to us today that it did not need quoting by name. It was first recognized by the pioneering geologist, Niels Steno (1638–1686). Simply stated, if we regard a sequence of sedimentary rocks, the oldest bed is at the bottom and the youngest at the top. That is, sedimentation is under the influence of gravity and beds characteristically accumulate from the base up.

Is there any way for beds (in the broadest sense) to accumulate out of order? The most likely such discordant intercalation is if a layer of molten igneous magma is intruded into, and forced between, two sedimentary beds. In theory such a structure, called a sill, should be differentiated from a lava flow by its baked boundaries. A lava flow extruded over a sedimentary surface will be hot, probably over 1000ºC, and will 'cook' the sediment; this process is called contact metamorphism. The cooked rocks will be hard and splintery, and the lava will cool and solidify before any overlying sedimentary beds are deposited. Thus, only the basal contact will demonstrate contact metamorphism. In contrast, a sill will cook both upper and lower contacts.

Beds are deposited in one orientation and one order, but the Earth has a habit of folding them, breaking them along faults, and even turning them upright or even upside down. It is not my intention to explain why this happens – there are many good books that will explain Earth movements (tectonics) to you – but movements in the crust pose the interesting question, are these beds the right way up or have they been inverted? The way up is a key feature of sedimentary beds that you must determine. Fortunately, there are many phenomena that only occur one way round (Fig. 6.1). Tucker (2011, pp. 24–25) provided a succinct introduction to sedimentary features that will help you determine way up. Herein, it is fossils that are our concern. For example, trees grow up and their roots grow down; an upright fossil tree will show the way up. For more discussion of fossils as way-up structures, see Chapter10.

Figure 6.1 Sedimentary cross-bedding, a way-up structure, exposed in the walls of Morpeth station, Northumbria, north-east England (after Donovan, 2018, fig. 3C). The block shows the details of two sides. The lower, essentially horizontal and parallel beds were cross-cut when the upper sandstone beds were deposited; erosion could only occur downwards and so this block is the right way up. Note that the truncations of the underlying beds by this erosive surface are plainly visible. Scale in cm (left) and inches (right).

The geologic record

If you have read this book so far, then I assume that you have a good appreciation of stratigraphy and the ordering of strata (beds). The major subdivisions of the Earth's crust occur in a set order, determined by their geological history. The rocks that include common shelly fossils are termed the Phanerozoic (from Greek, meaning 'evident life') and were deposited during the past 540 million years or so. The oldest Phanerozoic rocks belong to the Cambrian period (Table 6.1). Older rocks, lumped together as the Precambrian, are poorly fossiliferous and were deposited during a span of four billion years. Phanerozoic beds of different ages are recognized on the basis of their contrasting fossil content, their biostratigraphy (see below). For example, and at a coarse scale, all trilobites are Palaeozoic, but particular species of trilobite are limited to narrow parts of this succession. If you can identify your trilobite, you can determine the stratigraphic position of the rocks from which it was collected. For a better appreciation of the stratigraphical

Table 6.1 Schematic depiction of the major subdivisions of the Phanerozoic timescale (redrawn and modified after eesc.columbia.edu/courses/v1001/phanerozoic2.html); dates of interval boundaries after Gradstein et al. (2004). The Carboniferous is further separated into the (lower) Mississippian and (overlying) Pennsylvanian.

23.03	Neogene	**CENOZOIC**
	Paleogene	
65.5		
	Cretaceous	**MESOZOIC**
145.5		
	Jurassic	
199.6		
	Triassic	
251.0		
	Permian	**PALAEOZOIC**
299.0		
	Carboniferous	
359.2		
	Devonian	
416.0		
	Silurian	
443.7		
	Ordovician	
488.3		
	Cambrian	
542.0		

record, I recommend Ager (1993) – perhaps a little long in the tooth, but still pithy and thought-provoking.

Angular unconformity

Consider the sedimentary bed. It is separated from the bed above and below it by bedding planes. A bed represents the time necessary for that thickness of sediment to be deposited; compaction and cementation came later. But what does the bedding plane represent? Obviously, it is a gap in deposition and change in sedimentation.

Let me introduce one more significant physical, rather than palaeontological, feature of stratigraphic geology. Nowhere is there a complete succession of Phanerozoic strata. Breaks in sedimentation, analogous to gaps in time, may be small (bedding planes between beds) or large. The time periods represented by bedding planes are worthy of a philosophical treatise, but, to attempt to summarize a difficult concept in simple terms, in any sequence of discrete beds, it is certain that the bedding planes represent a greater extent of time than do the rocks themselves.

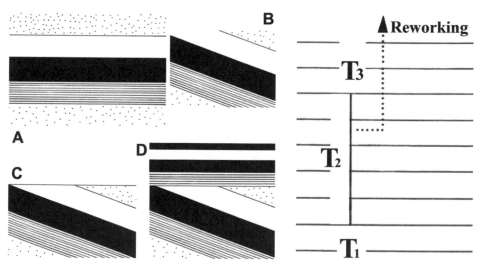

[Left] Figure 6.2 Schematic diagram of the formation of an angular unconformity. (A) Deposition of a sequence of sedimentary beds. (B) Earth movements tilt these beds. (C) Erosion planes off the tilted beds. (D) A new sequence of beds is deposited on the erosion surface, which is now the plane of angular unconformity.

[Right] Figure 6.3 Schematic diagram to illustrate some theory of biostratigraphy (redrawn and modified after Paul, 1985, fig. 3). The vertical axis represents time; the species is represented by the vertical line and represented by a time T_2 from its evolution to its extinction. All the time before the evolution of the species is T_1; all the time after its demise is T_3. Time T_2 is thus represented by a limited stratigraphic interval. The only possible confusion will occur if this species is reworked, eroded out of beds of equivalent to age T_2 and redeposited in beds equivalent to age T_3 (see Chapter 9).

If even greater lengths of past time are invoked, then the beds may be uplifted, tilted and planed off by mountain-building episodes (Fig. 6.2). New beds deposited horizontally above such deformed beds will overlie a stark division, called an angular unconformity. Angular unconformities are best seen in quarry or cliff sections and commonly represent long gaps of geological time, perhaps tens or hundreds of millions of years.

Biostratigraphy

Back to palaeontology. How is our understanding of fossils able to subdivide the rock sequences of the Phanerozoic? The essence is that each species divides geological time into three unequal segments (Paul, 1985). There was a vast period of time after the origin of the Earth, about 4.6 billion years ago, before a given Phanerozoic species evolved: let us call this length of time T_1 (Fig. 6.3). The species then existed for a length of time, T_2, from the evolution of the first individual to the extinction of the last. This is followed by time T_3 after the demise of the species, never to be seen again. One species does this, so consider the complexity of the pattern that tens or hundreds of species in time might generate. Essentially, they subdivide Phanerozoic time into a unique pattern of species durations and co-occurrences. Formal subdivisions based on species distributions are called biozones, or zones for short, each defined by a named species – the zone fossil – but this is inevitably associated with many others. Each has a unique stratigraphic range and provides individual data.

References

Ager, D.V. (1993) *The Nature of the Stratigraphical Record*. Third edition. Wiley, Chichester.

Donovan, S.K. (2018) Using urban geology: a field guide to Morpeth Railway Station, northern England. *Geology Today*, **34**: 97–99.

Gradstein, F., Ogg, J. & Smith, A. (2004) *A Geologic Time Scale 2004*. Cambridge University Press, Cambridge.

Paul, C.R.C. (1985) The adequacy of the fossil record reconsidered. *Special Papers in Palaeontology*, **33**: 7–15.

Tucker, M.E. (2011) *Sedimentary Rocks in the Field: A Practical Guide*. Fourth edition. Wiley-Blackwell, Chichester.

CHAPTER 7

FOSSILIFEROUS SEDIMENTARY ROCKS: SILICICLASTICS

The siliciclastic sedimentary rocks are mainly the result of mechanical breakdown of pre-existing, more or less silica-rich parent rocks. Quartz, SiO_2, is the commonest mineral in the Earth crust because it is physically tough (7 on Mohs' scale of hardness) and chemically stable (almost insoluble in water). It reaches the surface in silica-rich igneous rocks, such as rhyolites (extrusive, that is, lava) and granites (intrusive, molten bodies that cool and solidify in the Earth's crust). These are chemically stable at higher temperatures and pressures than are found at the surface of the crust, and slowly break down both chemically and physically. Quartz grains are released, while other silicate minerals, such as feldspars and micas, are broken down chemically to form, eventually, clay minerals.

Siliciclastic sedimentary rocks are classified on the basis of grain size. A conglomerate may include boulders as big as a car (or bigger), whereas those in a mudrock are so tiny as to be invisible to the naked eye (Table 7.1). Conglomerates are indicative of a

Rock type	Grain size
Boulder conglomerate/breccia	Greater than 256 mm
Cobble conglomerate/breccia	64–256 mm
Pebble conglomerate/breccia	4–64 mm
Gritstone	2–4 mm
Sandstone: coarse-grained	½ to 2 mm
Sandstone: medium-grained	¼ to ½ mm
Sandstone: fine-grained	1/16 to ¼ mm
Siltstone	1/256 to 1/16 mm
Mudrocks (shale, mudstone)	Less than 1/256 mm

Table 7.1 A grain-size classification of clastic sedimentary rocks, compiled from various sources (including Maley, 2005, p. 19). Most grain sizes can be determined in the field with the naked eye and hand lens. Siltstones and mudrocks are differentiated in the field by rubbing them, gently, over your teeth; mudrocks will be smooth, but siltstones will grate (but not too much) due to their fractionally larger grains, some of which may be apparent with a hand lens. Shales are fissile whereas mudstones are more massive. A conglomerate has more or less rounded grains; a breccia is conglomerate-like with angular grains.

particularly high-energy setting in which pebbles, cobbles and/or boulders, collectively termed clasts (smaller sedimentary fragments are grains), were transported and became rounded by abrasion. A breccia contains angular clasts, whereas those of a conglomerate are rounded.

Conglomerates may be fossiliferous, but the fossils are more likely to be found in the clasts; that is, they occur in the pebbles, cobbles and boulders derived from the rock that was eroded to make the conglomerate: in which case, beware.

When I taught at the University of the West Indies in Jamaica, I led an annual fieldtrip for first year students to Farquhar's Beach on the south coast. The rocks in this section showed a huge range of sedimentary and tectonic features, the best teaching locality that I know (see also Chapter 9). An angular unconformity is well exposed over many hundreds of metres, and huge boulders of the Pleistocene 'red bed' conglomerate from above it litter the beach; some are as big as a garden shed or garage. I would ask the students what environment was represented. They would look at the limestone cobbles and boulders in the conglomerate, and correctly identify massive scleractinian corals, sand dollars (= echinoids) and large gastropods. Marine, surely? Well, no. The fossiliferous clasts came from the source rock, a Miocene marine limestone. The conglomerate itself is terrestrial. Apart from being a 'red bed', fossils in the matrix between the clasts include root casts and land snails (Donovan & Paul, 2013).

Yet, despite such a good example, conglomerates are rarely fossiliferous, certainly more rarely than sandstones, siltstones and mudrocks. This is explained by grain size and energy/environment of deposition. A current flow capable of carrying pebbles and larger grains is equally likely to grind shells down to small, unidentifiable fragments. A coarse-grained sandstone is still indicative of high-energy deposition, but some shells, at least, will be much larger than individual sand grains and, in consequence, relatively more durable.

A feature of sandstones and other finer-grained siliciclastics is that fossils may be preserved as natural moulds and casts. Essentially, these rocks may be porous (spaces in the rock) and permeable, the pores allowing the passage of percolating groundwaters. These will likely be weakly acidic due to dissolved carbon dioxide,and would react to dissolve carbonate shells. After solution, the resulting cavity may be infilled by different minerals precipitating from groundwaters. Thus, an originally calcite shell may be completely dissolved and then the resulting cavity infilled by, for example, calcite (again), silica (SiO_2) or pyrite (FeS_2) (Fig. 21.1).

My Ph.D. research was largely concerned with such 'dissolved away' fossils. They present problems to the observer of perspective. The fossil is not seen as its body, but as an impression; compare it with, for example, your foot and the impression it makes in the sand on the beach. Just as a fictional detective might cast a footprint in plaster, so a palaeontologist should cast a (robust) natural mould. This will give you a cast of how the fossil actually appeared in life. The practical side of casting is discussed in Chapter 32.

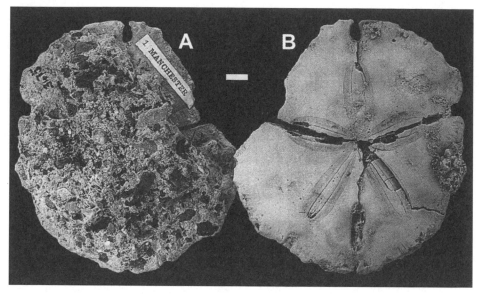

Figure 7.1 The sand dollar (echinoid) *Encope homala* Arnold & Clark, holotype, Museum of Comparative Zoology 3475 (after Donovan *et al.*, 1994, fig. 1). **(A)** Oral (basal) surface showing pebble conglomeratic nature of enclosing sedimentary rock. No sand dollar could live in such a coarse-grained sediment, so the specimen must have been transported. **(B)** Apical surface (whitened with ammonium chloride). Scale bar represents 10 mm.

Fossils in sandstones and conglomerates need to be robust enough to survive any transport before final burial. An environment that is energetic enough to roll pebbles and sand grains along, until the energy dissipates and they are 'dropped' in a lower energy setting, will also transport shells. Figure 7.1 illustrates the sand dollar *Encope homala* Arnold & Clark. The species was originally described on the basis of this specimen only; we now have many more. The defining reference for any named species is the specimen(s) from which it was originally described. These are called types. If described from a single specimen, it is called the holotype. If described from two or more specimens, one is designated holotype, which is the master reference, and the others are paratypes.

As originally published, only the apical surface of the holotype *E. homala* was illustrated (Fig. 7.1B) with the cryptic locality parish of Manchester, Jamaica. This is like saying a fossil came from Lancashire, Noord Holland or Ohio; poorly localized, indeed. So, I went to see the specimen in the Museum of Comparative Zoology in Harvard, Massachusetts, and I turned it over (Fig. 7.1A). Eureka! It was preserved in a pebble conglomerate. In the parish of Manchester, this had to be a coastal exposure, as most of the inland rock outcrop is limestone. And we found the site (Donovan *et al.*, 1994).

Siltstones and mudrocks – that is, shales and mudstones – are the finer-grained silici-clastic rocks and, in consequence, are always likely to preserve specimens particularly well. There are two reasons for this: fine-grained indicates low energy, so specimens will not be damaged by high-energy transport; and fine grains will favour a high fidelity of preservation. Some of the best units for collecting are mud-rich formations of the Mesozoic and Cenozoic, exposed along the south coast of mainland Britain, from Kent to Dorset. These rocks demand respect, commonly forming high cliff lines that are characteristically unstable. When dry, they are prone to landslides, both big and small; when wet, to mudslides; and, commonly, to mongrel combinations of the two. They will be particularly unstable in the spring, after winter freezes have loosened parts of the face by frost wedging.

But these rocks can be collected without risk to life and limb. Stay away from the cliffs. The mudrocks are broken down by the action of the sea, destroying some fossils, but washing out many others, which accumulate and concentrate on the beach. Context is lost, but preservation may be excellent. So, again, there are advantages in resorting to beachcombing. Look out for fossils in the round, but also watch for nodules, diagenetic hardening of rocks that are commonly formed in the altered geochemical environment around a rotting organism, preserving the hard shell within. Nodules resist compression of mudrocks due to weight of overburden. Thus, a fossil in a nodule is commonly better preserved than the same species in the surrounding mudrocks.

References

Donovan, S.K., Dixon, H.L., Littlewood, D.T.J., Milsom, C.V. & Norman, Y.J.C. (1994) The clypeasteroid echinoid *Encope homala* Arnold and Clark, 1934, in the Cenozoic of Jamaica. *Caribbean Journal of Science*, **30**: 171–180.

Donovan, S.K. & Paul, C.R.C. (2013) Late Pleistocene land snails from 'red bed' deposits, Round Hill, south central Jamaica. *Alcheringa*, **37**: 273–284.

Maley, T.S. (2005) *Field Geology Illustrated*. Second edition. Mineral Land Publications, Boise, Idaho.

CHAPTER 8

FOSSILIFEROUS SEDIMENTARY ROCKS: LIMESTONES, FLINTS, CHERTS AND COALS

What could be more obvious than fossils with a lime (calcium carbonate) shell preserved in limestones? Yet just because a fossil is preserved there does not make it easy to collect. Indeed, shells in well-lithified limestones are a pig to collect, nothing less. The lack of mineralogical distinction between a calcium carbonate shell and a limestone means that they are likely to behave as one. In many cases the fossil cannot be removed from the rock without time-consuming mechanical development using a pin in a pin vice or similar apparatus. There are two simple solutions to such a dilemma: either accept the fossil as it is preserved; or look for less well-lithified horizons in the succession.

Collecting from limestones is a personal favourite of mine. Some limestones are interbedded with more muddy, less well-lithified horizons. This can be removed with relative ease to expose the specimens preserved at the limestone surface. Thus, the famous fossils of the Much Wenlock Limestone Formation at Dudley, West Midlands, are well known for their complete preservation (Fig. 8.1A), but the specimens are only so well known because they can easily be washed clean. In contrast are the Carboniferous limestones used as building facing stones, where the fossils are only seen in section (Fig. 8.1B). It is the dominance of such massive limestones in areas like the White Peak of Derbyshire that has restricted description of the very common crinoid and echinoid remains preserved therein (Donovan, 2013).

So, fossils may be locally common in limestones, but they can be difficult to collect. One of the reasons that the Upper Cretaceous chalk is so popular with collectors is not that the fossils are always common – they can be particularly sparse – but that it may 'give up' its fossils with relative ease. This explains the attraction of beachcombing for fossils on chalk coasts; specimens may be complete or nearly so, just sitting on the beach and waiting for you.

This brings us to flint in particular and cherts in general. Flint in chalk is a secondary deposit, formed by mobile silica in alkali solution, which is precipitated along joints and beds. In other limestones, mineralization of identical composition is called chert.

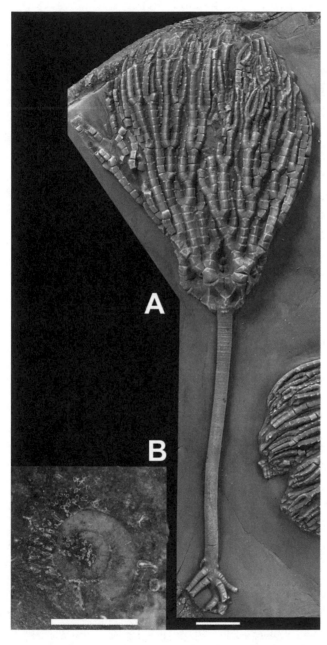

Figure 8.1 (A) The Silurian crinoid *Gissocrinus goniodactylus* (Phillips), Natural History Museum, London, specimen E45528, from the Much Wenlock Limestone Formation, Dudley, Worcestershire (after Donovan *et al.*, 2008, fig. 8C). Scale bar represents 10 mm. (B) The Mississippian gastropod *Straparollus*(?) sp. in a paving slab in Leiden, the Netherlands, encrusted by the tabulate coral *Syringopora* sp. (after Donovan, 2016, fig. 1). Scale bar represents 50 mm.

Flints and cherts may contain fossils. In particular, the silica of flint precipitates around siliceous sponges, but they can also be exotic, infilling trace fossils, and forming internal (called *Steinkerns*) and external moulds (Fig. 47.2C, D).

My favourite cherts are in Carboniferous limestones and rich in crinoid debris. Some crinoids are replaced by silica, but the calcite of others is dissolved out with time, leaving behind natural moulds in the chert. A pluricolumnal with an infill to its axial canal has the appearance of a screw—hence the name of these rocks as screwstones or, in the White Peak, Derbyshire screws.

Fossils in flints and cherts offer the possibility of being easily 'cleaned' by weak mineral acid, such as 10% hydrochloric acid. Any residual calcite of the fossils will be dissolved away, leaving a pristine mould. This is rarely necessary, but some specimens are part siliceous mould and part calcite remnant of the fossils, so they are neither one thing nor the other.

Grinding down a limestone surface will expose the included fossils to better effect, but in two dimensions. The freshly polished facing stones of new buildings give a good idea of this effect. Older stone buildings will be etched, the result of tens or hundreds of years of acid rain. A polished slab in your collection and stored indoors will remain pristine.

If you are going to polish a limestone to produce a clean, planar surface, you will need a glass plate, fine- and coarse-grained grinding powder, running water and elbow grease. Some sort of rotating wheel can make life easy, but good results may be obtained using a sheet of glass. Thinly spread coarse-grained powder on the plate, moisten it, then move your rock with a circular motion, around and around and – to see what progress is being made, wash the paste off the rock. A time arrives when there seems to be no further change. Wash the coarse-grained paste off the rock and off the plate thoroughly – you do not want to mix the coarse-grained with the fine-grained powder. Thinly spread fine-grained powder on the plate, moisten it, then move your rock around and around, wash and repeat until satisfied. Wash thoroughly and allow to dry. The limestone will always look better when wet.

Just as siliciclastic rocks (Chapter 7) are formed from fragments of pre-existing rocks cemented together, so limestones and coals should be thought of as chemical rocks, formed from organisms that, in life, concentrate carbon-rich components from the environment. Limestones are commonly, but not invariably, formed from fragmentary skeletons of invertebrates, from the macro- to the microscopic. Coals are formed under conditions in which dead plant organs are trapped in conditions where oxygen is excluded, becoming preserved as carbon films and woody structures.

Most fossiliferous deposits considered herein were marine in origin. Coals are terrestrial or marginal terrestrial deposits, perhaps interbedded with marine deposits that, together, are indicative of ancient oscillations of sea level. Like limestones, coals can be seen as chemical deposits, dominantly carbon. Coals form by processes that are analogous to modern vacuum-packing techniques. Consider a leaf that falls off a tree in autumn: over several days, it dries out, turns brown, wrinkles, becomes brittle and eventually breaks up into ever-smaller fragments.

This is not the way to make a coal. Rather, a coal-forming leaf (or other plant organ – branch, root, trunk or whatever) needs to enter an oxygen-depleted environment shortly after death. For example, the leaf sinks into an anoxic pond or lake in which plant debris builds up to a thickness until it is crushed by an overburden of muddy sediment and flattened. A common estimate of this flattening is ~15:1. That is, 15 m of plants make 1 m of coal. The dominant fossils of coals are plants, but there will be associated, minor invertebrates such as associated insects, perhaps more easily detected in interbedded mudrocks.

References

Donovan, S.K. (2012) Taphonomy and significance of rare Chalk (Late Cretaceous) echinoderms preserved as beach clasts, north Norfolk, UK. *Proceedings of the Yorkshire Geological Society*, **59**: 109–113.

Donovan, S.K. (2013) Where are all the crinoids? An enigma of the Lower Carboniferous (Mississippian) White Peak of midland England. *Geology Today*, **29**: 108–112.

Donovan, S.K. (2016) A mollusc–coral interaction in a paving slab, Leiden, the Netherlands. *Bulletin of the Mizunami Fossil Museum*, **42**: 45–46.

Donovan, S.K., Lewis, D.N., Crabb, P. & Widdison, R.E. (2008) A field guide to the Silurian Echinodermata of the British Isles: Part 2 – Crinoidea, minor groups and discussion. *Proceedings of the Yorkshire Geological Society*, **57**. 29–60.

CHAPTER 9

REWORKED FOSSILS

Reworked *Kuphus* tubes and their borings

I am going to introduce something of the nature and interest of reworked fossils (Fig. 6.3) by reference to a specific example. The late Pleistocene 'red bed' conglomerates that are exposed above an angular unconformity on Jamaica's central south coast are fascinating deposits that remain understudied. I am particularly interested in the land snails, moderately varied, but their diversity inadequately determined so far (Donovan & Paul, 2013). The conglomerate clasts are boulders of Miocene limestone, reworked out of nearby Round Hill, an adjacent mass of the Newport Formation (see also Chapter 7). Thus, within a Pleistocene conglomerate are clasts of Miocene limestone, containing fossils. These are referred to as reworked fossils, removed from their host rock by erosion and redeposited in a younger deposit.

But look more closely. At the base of these red beds, seen close to beach level in the eastern part of the exposure, are fragments of *Kuphus* tubes. *Kuphus* is a crypt-dwelling bivalve. In life, it burrows into the sediment, but secretes an enclosing, and robust, calcareous tube to the surface. The siphons of the bivalve need to extend the length of the tube to the surface, to access water rich in microscopic food and oxygenated water. These *Kuphus* tubes are similarly reworked as fragments from the Newport Formation, but rare specimens have been bored (Fig. 9.1).

Borings in reworked fossils are unusual. The tubes of *Kuphus* are not a trace fossil, but part of the body fossil of an encrypted bivalve. In the Oligo-Miocene limestones of Jamaica, *Kuphus* is commonly preserved perpendicular to bedding, that is, in life position. Borings would not have been possible in such tubes, which were embedded in the sediment both in life (Fig. 9.2A) and after (Fig. 9.2B). No borings have been recognized in *Kuphus* in the Oligo-Miocene of Jamaica. Thus, the borings in *Kuphus* in the Farquhar's Beach red beds are probably younger.

The canals of the *Kuphus* crypts are invariably infilled with well-lithified limestone in intimate association with the internal walls. In contrast, all borings in *Kuphus* crypts are open and preserve no suggestion that they have ever been infilled. Again, this suggests that the borings are post-Miocene.

A post-Miocene age of these borings is further supported by the evidence of provenance. RGM 544 453 (Fig. 9.1C) was removed from the basal bed of cobbles rich with

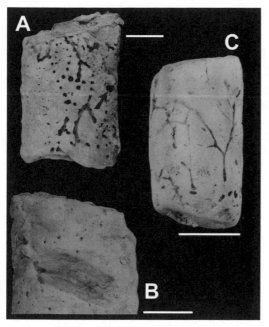

Figure 9.1 Reworked *Kuphus* sp. tubes, Farquhar's Beach, parish of Clarendon, Jamaica (after Donovan *et al.*, 2010, fig. 3). (**A, B**) Naturalis Biodiversity Center, Leiden, RGM 544 452. (**A**) Side of tube with numerous *Entobia* isp. borings and one *Gastrochaenolites* isp. (arrowed), produced by clionaid sponges and boring bivalves, respectively. (**B**) Detail of reverse of tube showing oblique groove referred to *Gastrochaenolites*? isp. (**C**) RGM 544 453, tube with numerous *Entobia* isp. Scale bars represent 10 mm.

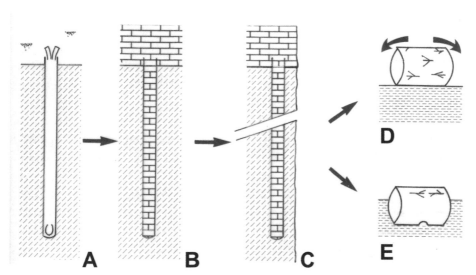

Figure 9.2 Schematic history of bored *Kuphus* tubes in the Miocene (**A, B**) and Late Pleistocene of south central Jamaica (**C–E**) (after Donovan *et al.*, 2010, fig. 4). (**A**) Living bivalve secretes *Kuphus* crypt (tube); note siphons of living bivalve extending above sediment surface and shell near base of tube. (**B**) Dead *Kuphus* tube infilled by lime sediment and lithified. (**C**) Fossil *Kuphus* tube becoming disinterred (rock face to right); Pleistocene bivalve boring cross cuts one side of the tube. (**D**) RGM 544 453, disinterred tube fragment rolls around on Pleistocene sea floor, is infested by boring sponges (*Entobia* isp.) and is later re-buried. (**E**) RGM 544 452, disinterred tube fragment more or less stable on Pleistocene sea floor, is infested by boring sponges and a boring bivalve (*Gastrochaenolites* isp.; not shown) mainly on one side, and is later completely re-buried.

the boring *Gastrochaenolites* isp. Thus, although only one specimen was collected *in situ*, it was found in a horizon rich in trace fossils of marine origin.

Borings in *Kuphus* crypts are probably only possible after exhumation from the White Limestone Group in the Late Pleistocene (Fig. 9.2C). This association is not simple; it has an unusual longevity. The Miocene *Kuphus* were eroded out of Round Hill and formed pebbles on the late Pleistocene sea floor (Fig. 9.2D, E); RGM 544 453 was collected *in situ* from the basal Farquhar's Beach red beds. Most of these red beds were terrestrial, but the basal part of this sequence, immediately above the angular unconformity, was shallow marine in origin. The *Kuphus* tubes, sporting borings typically produced by marine sponges and bivalves, support this interpretation. Thus, Miocene *Kuphus* were infested by borers after a gap of many millions of years.

The importance of reworked fossils

What is the geological interest and importance of reworked fossils? In part, they need to be recognized because of the possibility that they could distort our concepts of biostratigraphy and time. From your own experience, think of the fossils that you may have collected from the beach: these have been reworked out of an old deposit and now have the potential to be reworked into a new, younger bed. For example, Mississippian corals at Cleveleys (Chapter 48) were reworked into Pleistocene 'boulder clays' and then reworked a second time to become beach clasts. They are, if you like, clasts in waiting for the next deposit.

There are two broad styles of reworking. Being reworked inside a lithified clast will provide protection to even a fragile fossil; the branching corals at Cleveleys would not have been preserved in three dimensions without being embedded in lithified limestone (Fig. 48.4).

Alternately, some fossils are particularly robust and survive reworking even when released from the rock. For example, you have already been introduced to the tough tubes of *Kuphus*, not preserved complete (which may be >1 m long), yet easily identified. The Upper Cretaceous echinoid *Echinocorys scutata* Leske is moderately common as a beach clast between Overstrand and Cromer on the north Norfolk coast (Chapter 47). Again returning to Jamaica, Cretaceous rudist bivalves may occur in Eocene siliciclastic successions (Pickerill *et al.*, 1995). Indeed, one species of Jamaican Cretaceous rudist is known only as a reworked fossil in the Eocene succession.

References

Donovan, S.K., Blissett, D.J. & Jackson, T.A. (2010) Reworked fossils, ichnology and palaeoecology: an example from the Neogene of Jamaica. *Lethaia*, **43**: 441–444.

Donovan, S.K. & Paul, C.R.C. (2013) Late Pleistocene land snails from 'red bed' deposits, Round Hill, south central Jamaica. *Alcheringa*, **37**: 273–284.

Pickerill, R.K., Donovan, S.K. & Dunn, J.T. (1995) Enigmatic cobbles and boulders in the Paleogene Richmond Formation of eastern Jamaica. *Caribbean Journal of Science*, **31**: 185–199.

CHAPTER 10

FOSSILS AS WAY-UP STRUCTURES

The concept of way-up structures was introduced in Chapter 6. In the bedded sedimentary strata in which we are interested, there are many potential indicators of way-up. Many are physical structures preserved in sedimentary rocks (en.wikibooks.org/wiki/Historical_Geology/way-up structures). To give but one example as an illustration, consider a bed of sandstone in which the grains are graded, with coarse grains at the bottom and fining upwards. With only rare exceptions (I have never seen one), the coarse-to-fine-grained transition indicates the base-to-top of a bed, that is, way-up, with the heaviest grains deposited first by a waning water current. Even a volcanic lava flow can show way-up: the underlying beds will be 'cooked' and splintery; the top may be rubbly; and frozen gas bubbles in the flow will get bigger towards the top. The last can be modelled easily – just watch the bubbles combine and grow the next time that you unscrew the stopper of a carbonated drink.

But what of fossils? I would suggest that there are three principal way-up structures that we can seek in fossils: their life orientation; their hydrodynamic stability (also see Chapter 11); and geopetal infills. There may be many more, some not defined herein and others that have yet to be recognized – keep your eyes peeled for such unknowns.

Life orientation

If you are confident that your benthic fossil has not been transported and is preserved in its life position, then what can it tell you about way-up? Be careful – if you have one fossil that seems to be undisturbed it may tell you something, but 50 specimens preserved in the same orientation will be more likely to convey a true signal.

Obviously our best way-up indicators are likely to be members of the sessile benthos, those organisms that spent their lives sitting on the seafloor. The most likely common indicators are massive: they will not have been disturbed by anything but a major event, such as a storm current. Their form is such that most currents will probably have flowed over and around them (Fig. 10.1). Yet remember that they can be moved and transported – a bed with domed corals in other orientations than in life (consider a storm beach in

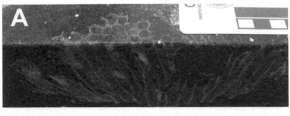

Figure 10.1 Section of a massive colony of the Mississippian tabulate coral genus *Michelinia* sp. (after Van Ruiten & Donovan, 2018, fig. 2C, D), presumed in life orientation. Two views of a colony found in a windowsill of the University of Leiden at Rapenburg 71. (**A**) Oblique view. (**B**) Longitudinal section. Scale in cm. Also see Figure 45.2.

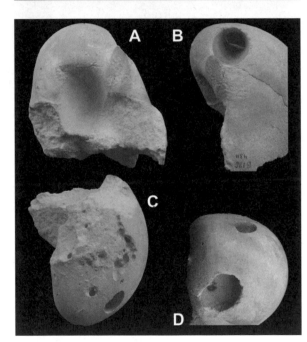

Figure 10.2 The club-shaped boring *Gastrochaenolites* isp. in an incomplete internal mould of an Upper Cretaceous nautiloid (Natuurhistoruisch Museum Maastricht IndeBraekt Collection, no. 4311), from the Meerssen Member (Maastricht Formation) at the former Ankerpoort-Curfs quarry, Geulhem, the Netherlands (after Donovan & Jagt, 2018, fig. 4). (**A**) Septal view of the youngest phragmocone chamber, borings not apparent. (**B–D**) *Gastrochaenolites* isp. seen in oblique-lateral (**B**), lateral (**C**) and ventral (**D**) aspects. (**D**) is presumably the approximate orientation of the nautiloid when bored. The greatest whorl width (**A**) is 93 mm.

the tropics, littered with coral debris derived from the reef offshore) will tell its own story of disruption and transport.

Features of hardground surfaces include good way-up indicators such as borings (going down) and diverse encrusters on the surface. But be cautious – both can be found in transported clasts (Fig. 10.2). So, again beware. Further, cavities in hardgrounds may be encrusted. The way-up structures, such as encrusting acorn barnacles and oysters, may be upside down because they are attached to the roof of the cavity although, overall, the bed may be the right way up.

Hydrodynamic stability

Treat yourself to a walk on the beach, preferably one with an abundance of disarticulated bivalve molluscs. The sun shines, the seagulls swoop and the bivalves preserve a
signal. They will commonly, albeit not invariably, be preserved concave down. That is,
each is a little domed 'hump' on the beach. The reason for this will be easy to see if you
watch their interactions with the waves – it is a stable orientation, each valve acting as a
tiny ripple over which the water flows with minimum disturbance. A bed of disarticulated fossil shells, bivalves or brachiopods, can provide similar information.

A little less common than fossil bivalves are irregular echinoids. Some irregulars
are commonly rather more durable than most regular echinoids, having plates that
are well-sutured together. Robust irregular echinoids can thus be transported like
bivalves and form echinoid-rich horizons, such as the *Echinocorys* band at Margate
on the north Kent coast (Donovan, 2020). Upper Cretaceous *Echinocorys* Leske and
related taxa such as *Hemipneustes* L. Agassiz were members of the holasteroids, which
lived as surface furrowers. They have a flat bottom and are commonly preserved
with this surface in contact with the sediment. Possibly the toughest echinoids were,
and are, the Cenozoic clypeasteroids or sand dollars, commonly shallow burrowers.
Again, while they are solid enough to be reworked, they are commonly preserved
with the oral surface down and the apex forming a low, ripple-like upper surface (see
Chapter 7).

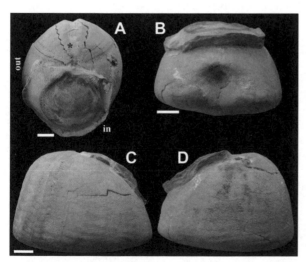

Figure 10.3 The Upper Cretaceous echinoid *Hemipneustes striatoradiatus* (Leske) encrusted by the left (= attached) valve of the oyster *Pycnodonte* (*Phygraea*) *vesiculare* (Lamarck) (after Donovan & Jagt, fig. 2). From the Nekum Member (Maastricht Formation) of quarry 't Rooth, Bemelen (southern Limburg, the Netherlands; Natuurhistoruisch Museum Maastricht MK.151). **(A)** Apical view, anterior towards top of page. *Pycnodonte* (*Ph.*) *vesiculare* attached to posterior of apical system and much of the posterior test. The long, straight dorsal margin is arrayed sub-perpendicular to the long axis of the echinoid. A small black asterisk marks to positions of (pale-coloured) small oysters in the anterior ambulacrum. 'In' and 'out' mark the approximate incurrent and outcurrent positions of the oyster. **(B)** Posterior view showing how *Pycnodonte* (*Ph.*) *vesiculare* extended to just above the periproct. **(C, D)** Left and right lateral views, respectively, showing *Pycnodonte* (*Ph.*) *vesiculare* extending across the posterior test. Stated orientations refer to the echinoid test. All scale bars represent 10 mm.

Holasteroids are high and clypeasteroids mostly low, but, like bivalves, both are commonly preserved as a 'ripple' with their flat base in contact with the seafloor and their domed apical surface uppermost. Further, these echinoids may form benthic islands, hard substrates on a sandy or muddy seafloor. They can be favourites for boring and encrusting organisms, which infest the domed apical surface, but not the basal surface in touch with the seafloor (Fig. 10.3).

Geopetal infills

Bedding can lie. As an undergraduate I went on a fieldtrip led by my palaeontology lecturer, Dr Fred Broadhurst, to the Mississippian reef of the Castleton area, Derbyshire. We stopped at a site above the Treak Cliff Cavern and looked at a steeply dipping bed. Obviously it had been uplifted and tilted, or was it something more subtle? The latter, in fact. Fred demonstrated a fossil in the said bed which indicated deposition of the steep bed on the front of the reef: that is, it was laid down sloping down the front of the reef. What was the evidence? A geopetal infill (Fig. 10.4).

Geopetal infills are called palaeo-spirit levels by some – a good summing up of their properties. A hollow shell is buried and partly infilled with sediment. The top of the sediment surface is horizontal. With time, the sediment is lithified with a hollow space above it. This cavity may persist or be infilled by mineral crystals that grow from percolating groundwaters. The surface of the lithified rock infill is your spirit level – it will show the horizontal at the time of deposition. The more geopetal infills you find in a bed, the greater your confidence will be as to the original way-up.

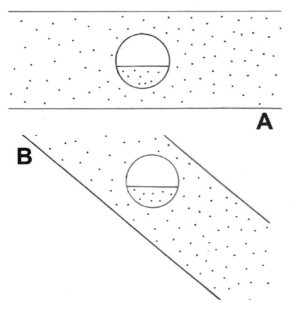

Figure 10.4 Geopetal infills. **(A)** The most likely preservational association. This horizontal bed preserves a globular shell incompletely filled with lithified sedimentary rock (= geopetal infill). The flat, horizontal top of the infill in the shell is parallel to the top and bottom of the bed, which was deposited horizontally. **(B)** An unusual situation (see text). On a reef front a bed was deposited on a slope. The horizontal top of the geopetal infill indicates that the enclosing bed was not deposited horizontally.

References

Donovan, S.K. (2020) Fossils explained 78. Never bored by borings: an ichnologist in Margate. *Geology Today*, **36**: 232-235.

Donovan, S.K. & Jagt, J.W.M. (2013) Aspects of clavate borings in the type Maastrichtian (Upper Cretaceous) of the Netherlands and Belgium. *Netherlands Journal of Geosciences*, **92**: 133–143.

Donovan, S.K. & Jagt, J.W.M. (2018) Big oyster, robust echinoid: an unusual association from the Maastrichtian type area (province of Limburg, southern Netherlands). *Swiss Journal of Palaeontology*, **137**: 357–361.

Van Ruiten, D.M. & Donovan, S.K. (2018) Provenance, systematics and palaeoecology of Mississippian (Lower Carboniferous) corals (subclasses Rugosa, Tabulata) preserved in an urban environment, Leiden, the Netherlands. *Bulletin of the Mizunami Fossil Museum*, **44**: 39–50.

CHAPTER 11

FOSSILS AS CURRENT INDICATORS

In palaeoecology there are many parameters that cannot be measured in contrast to modern ecology. An ecologist can sit in a field or on the seashore and watch things happen. Is it sunny or cloudy? Wet or dry? Hot or cold? Measurements may be fine-tuned. In palaeoecology, even such coarse observations are just not possible; the moment is long gone. Yet just because you cannot do something does not mean that you cannot do anything. An observation is always important and many are missed; how many ecologists are sitting in a field right now and determining if it is wet, hot or whatever?

More discussion of palaeoecology will come later (Chapters 18–20), but here I discuss one of the aspects related to sedimentology. So, what can we measure? Lots of things. In this chapter the focus is on how fossils can be used to determine a feature of the ancient environment, namely the direction of flow of current. One common way of recognizing ancient currents is to examine the orientations of long, thin fossils such as belemnites, crinoid stems or orthoconic nautiloids (Ager, 1963, p. 77 *et seq.*, pl. 3, fig. B).

Transported crinoid stems

This account is adapted from Donovan (2012). The slab of sandstone described herein (Figs 11.1, 11.2) preserves a monospecific accumulation of crinoid stems, some extending diagonally across the width of the slab and only truncated, artificially, at the edges; the complete preserved specimens were longer, but the size of the collected sample was determined by the limits imposed by the intersections of joint planes. That this slab preserves a spectacular fossil accumulation is apparent at first glance; its palaeobiological, taphonomic and sedimentological signals are all apparent once the correct questions are asked of it.

The specimen is part of the R. B. Rickards Ph.D. thesis collection of the Sedgwick Museum, Cambridge. It is from the Ecker Secker Beck, River Rawthey, Howgill Fells, North Yorkshire, northern England, about [NGR SD 692 953], in the Cautley district (see Ordnance Survey 1:25,000 topographic sheet OL19, *Howgill Fells & Upper Eden Valley*). Brathay Flags, Lower Silurian (Wenlock).

Figure 11.1 Sedgwick Museum, Cambridge registration number (SM) TN 1979.1, crinoids, Silurian (Wenlock), Howgill Fells, northern England (after Donovan, 2012, fig. 1). Bedding surface preserving a monospecific assemblage of sub-parallel crinoid pluricolumnals. Field orientation of specimen is unknown, so alignment of photograph is arbitrary and named edges in text (upper, right, lower, left) determined only from this figure. Whitened with ammonium chloride. Scale bar represents 10 mm.

The specimen is a brick-sized slab of well-lithified, fine-grained sandstone, weathering brown, and with maximum dimensions of about 200 x 105 x 40 mm (Fig. 11.1). The reverse side to the figured face bears a luxuriant growth of moss, much of it still green, suggesting that Figure 1 shows the base of a bed; the sandstone appears to be subtly graded (= way-up structures; see Chapter 10) and agrees with this supposition. At the upper edge of the illustration, columns have become decalcified and stained with iron, indicating that this side was exposed to the elements immediately prior to collection. The figured, fossiliferous face is trapezoid, about 169 (upper) x 89 (right) x 169 (lower) x 105 mm (left). Angular measurements were made from the upper edge.

The surface preserves a monospecific assemblage of elongate, slender, parallel to sub-parallel crinoid pluricolumnals. These are angled to the upper edge at about 45°, although this increases to 60° towards the left (Fig. 11.1). Many specimens can be traced across the full width of the slab and are truncated at either end only by the limits of the surface. Some closely-spaced pluricolumnals are sub-parallel and overlap; most prominently, one wider specimen in the upper left (Fig. 11.2A) is overlapped by several more slender and parallel columns, producing a superficially spiral-like effect. One specimen in this area is preserved in a broadly zigzag pattern.

The crinoids are indeterminate systematically. The range of diameters of these pluricolumnals (best seen in Fig. 11.2A), may indicate either that multiple generations are

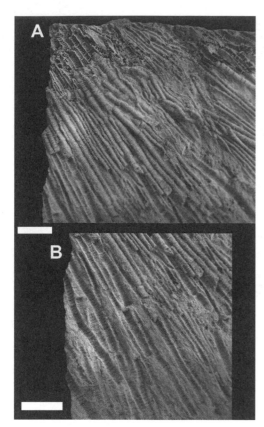

Figure 11.2 SM TN 1979.1, crinoid pluricolumnals, Silurian (Wenlock), Howgill Fells, northern England (after Donovan, 2012, fig.2). (**A**) 'Train-crash' crinoids *sensu* Broadhurst & Simpson (1973), slightly upper right of centre, column showing a sinuous pattern of disarticulation. Note the spiral effect of overlapping pluricolumnals towards the right. (**B**) Decalcified pluricolumnals (lower left in Fig. 11.1). Whitened with ammonium chloride. Scale bars represent 10 mm.

present (at least three?) or that the stem had both relatively narrower and broader regions. The former seems the more probable, implying that these specimens were derived from a true community of crinoids, albeit monospecific, with both younger and older individuals abounding.

The first obvious question about this unusual slab is where are the crowns and attachment structures? None are apparent, and it is simplest to assume that their absence is an accident of cut effect. That is, were these complete crinoids with the crowns and attachment structures intact, but not preserved on this slab because they were cut off by the small size of this bedding plane? This is at least possible, but not likely; the Silurian of the British Isles have yielded only rare, complete fossil crinoids. More probably, this depositional pattern was of long pluricolumnals only, disarticulated from both crowns and attachment structures. It is probable that adverse environmental conditions caused autotomy (self-mutilation) of the column rather than mechanical breakage. It is therefore suggested that transport has produced an allochthonous assemblage, although it may still be a life assemblage.

It is assumed that this interpretation is probable and that these long pluricolumnals were deposited by a single, energetic environmental event such as a turbidity current (an underwater 'landslide'). They obviously give some indication of current direction. At least one specimen is a 'train crash' crinoid pluricolumnal *sensu* Broadhurst & Simpson (1973, p. 372, fig. 4A; Donovan, 2021), the zigzag pattern suggesting disarticulation resulting from impact. Current direction was therefore parallel to the long axes of the pluricolumnals. The slight change in orientation towards the left (Fig. 11.1) may be indicative of local changes of flow direction or bed topography.

Swimming belemnites

The above example concerned dead organisms that were orientated by a current. But live organisms will also respond to a current and position their body accordingly. What if your fossil was moving around—what information might be preserved? A short, but significant, paper on current orientation in living belemnites and trace fossils in them was by Seilacher (1968). The radial calcite structure of the belemnite guard indicates that it must have grown enclosed by soft tissues, that is, it was within a squid-like body. If bored, they can be shown to be post-mortem in most examples (see, for example, Taylor *et al.*, 2013). Seilacher's specimens added further information to this model. His belemnite guards had been bored by acrothoracian barnacles, which are small and produce a distinctive sock-shaped pit. These had a consistent orientation and occurred throughout the entire circumference, through 360°. This consistent orientation indicates that the barnacles were in a constant current flow from one direction, but not lying on the sea floor. By comparison with extant acrothoracians, this flow was from the tip of the belemnite and along its length towards the body ('soft parts'). As it is unlikely that the belemnites were stationary in the water column, they must have been swimming as, in many extant cephalopods, by water jet propulsion blowing the animal along, tail first. It also demonstrates that in these specimens, at least, the soft tissues enclosing the guard must have been removed somehow, perhaps analogous to the way our own teeth erupt from the gums once fully grown.

References

Ager, D.V. (1963) *Principles of Paleoecology: An Introduction to the study of how and where Animals and Plants lived in the Past.* McGraw-Hill, New York.

Broadhurst, F.M. & Simpson, I.M. (1973) Bathymetry on a Carboniferous reef. *Lethaia*, **6**: 367–381.

Donovan, S.K. (2012) An unusual accumulation of crinoids from the Silurian of the Howgill Fells, Cumbria, UK. *Proceedings of the Yorkshire Geological Society*, **59**: 121–123.

Donovan, S.K. (2021). Train crash crinoids revisited. *Lethaia*, **54**: 1-3.

Seilacher, A. (1968) Swimming habits of belemnites—recorded by boring barnacles. *Palaeogeography, Palaeoclimatology, Palaeoecology*, **4**: 279–285.

Taylor, P.D., Barnbrook, J.A. & Sendino, C. (2013) Endolithic biota of belemnites from the Early Cretaceous Speeton Clay Formation of North Yorkshire, UK. *Proceedings of the Yorkshire Geological Society*, **59**: 227–245.

CHAPTER 12

YOUR PALAEONTOLOGICAL LIBRARY

If only you were here right now. I retire in October 2020 and will be leaving the Netherlands, going to live in the UK with Karen, my partner. Her apartment is a comfortable size, but does not have several spare rooms in which to store books and rocks. So, my library is being trimmed and trimmed hard. I am giving away over 95% of my geology books and papers. I have decided that if I am to research in my dotage, I will need to be focused, and intend to concentrate my efforts on trace fossils. But there are some references that I consider essential. These are my 'top twelve' as a guide to you as you start to accumulate publications on palaeontology. But new and worthy books are being published all the time (e.g., Wyse Jackson, 2019).

1, 2, 3: British Palaeozoic Fossils; British Mesozoic Fossils; British Cenozoic Fossils (Anon, 2012, 2013, 2017)

A week after I first went fossil collecting in 1975, I was in the British Museum (Natural History) and Geological Museum, as they then were, and bought the then-current editions of these books. Currently I have one set in the office and one at home. They are, without doubt, the most useful (and affordable) references to have to hand when you want to make a preliminary identification of that specimen, new to you, that you collected yesterday. The plates are of many excellent line drawings, arranged in stratigraphic order. Buy them new, or even a second-hand set of earlier editions should be grabbed as a bargain.

4: Invertebrate Palaeontology and Evolution (Clarkson, 1998)

My preferred fourth edition is a little long in the tooth—I have yet to see the new fifth edition, published in 2019—but this book is my first point of reference when I have a project involving an invertebrate fossil group with which I am not familiar. Euan Clarkson retired some years ago, since when there have been rumours that *Invertebrate Palaeontology* is being revised by two authors (never one) and not always the same two.

Euan obviously did the work of more than two lesser mortals who have failed to complete a new edition. The main groups that you are likely to encounter are dealt with in much more detail than merely a source of names for bits of skeletons.

5: Vertebrate Palaeontology (Benton, 2014)

I am not a vertebrate worker, apart from an odd dabble, but I read the first edition of Benton from cover to cover and found it an excellent guide for the uninitiated (like me). It is the vertebrate 'answer' to Clarkson and another book that has gone through multiple editions; remember, if a book is not highly successful, there is no need for more than one edition.

6: Palaeoecology: Ecosystems, Environments and Evolution (Brenchley & Harper, 1998)

This is a text that would benefit from a new edition, but, sadly, Pat Brenchley is no longer with us and Dave Harper is buried by other work. However, it still includes much that is relevant. A palaeoecological question is different from identifying a taxon – for example, what does this unusual feature or organism/organism association mean? These are questions we all ask, and the answer may not be easy; Brenchley & Harper will help you in the right direction.

7: Trace Fossils: Biology, Taphonomy and Applications (Bromley, 1996)

I seem to be recommending several books that are not so new, but do not despair: they are all good and full of meat. In the age of the internet, buying out-of-print books is both easy and, often, cheap. Bromley's *Trace Fossils* is a marvellous introduction to the subject, and can be read just for the sake of scientific entertainment, written by an author with a deep understanding of the ecology of the producing organisms.

8: The Origin of our Species (Stringer, 2011)

Fossil humans are rare and an unlikely pursuit for any collector. Yet there is an undoubted fascination in the subject; I have more than a metre of bookshelf filled on the subject, mainly by books on the collectors, the history of collecting and a pet subject, Piltdown Man. From among these, Stringer's book is a thoughtful introduction to *Homo sapiens*, its forebears and relations, and is here chosen as a book that gave me much pleasure in reading.

9: Wonderful Life: The Burgess Shale and the Nature of History (Gould, 1989)

Wonderful Life would be my choice for the finest book on natural history and the history of science of my lifetime. Gould has examined the early Palaeozoic evolutionary radiation by reference to its most famous manifestation, the Middle Cambrian Burgess Shale of British Columbia. The middle third of the book is systematics, yet it reads like a thriller. Our knowledge of these bizarre creatures has certainly progressed in the last 30 years; nonetheless, this is still the most satisfactory introduction to these enigmatic fossils.

10: The Great Fossil Enigma: The Search for the Conodont Animal (Knell, 2013)

Knell's volume is as well written and readable as *Wonderful Life*, but focuses on one group of enigmatic microfossils, the conodonts. These tooth-like elements are common fossils in Palaeozoic strata, particularly limestones, have been used by stratigraphers since the mid-nineteenth century, yet the (soft-bodied) animals that produced them have only been known since 1983. This history of their discovery and interpretation is a fascinating work of the history of science.

11: Sedimentary Rocks in the Field: A Practical Guide (Tucker, 2011)

I finish with two practical guides. Tucker's is an indispensable reference for the field palaeontologist. Part of our understanding of the occurrence and palaeoecology of any fossil must consider the information provided by the sedimentary rocks in which it is preserved. *Sedimentary Rocks in the Field* is small enough to slip into your pocket, and a book that should be dipped into for any and all answers to questions of sedimentary lore. I have been using this book since the first edition.

12: Writing for Earth Scientists: 52 Lessons in Academic Publishing (Donovan, 2017)

Apologies for this bit of self-promotion, but if you intend to write about fossils, then *Writing for Earth Scientists* is required reading. It is the only modern text specifically on this subject and, as the author is a palaeontologist, its 'slant' is in the direction of your studies.

References

Anon. (2012) *British Palaeozoic Fossils*. Revised by L.R.M. Cocks; Fifth edition. Natural History Museum, London.

Anon. (2013) *British Mesozoic Fossils*. Revised by A.B. Smith; Seventh edition. Natural History Museum, London.

Anon. (2017) *British Cenozoic Fossils*. Revised by J.A. Todd; Seventh edition. Natural History Museum, London.

Benton, M.J. (2014) *Vertebrate Palaeontology*. Fourth edition. Wiley, Chichester.

Brenchley, P.J. & Harper, D.A.T. (1998) *Palaeoecology: Ecosystems, Environments and Evolution*. Chapman & Hall, London.

Bromley, R.G. (1996) *Trace Fossils: Biology, Taphonomy and Applications*. Second edition. Chapman & Hall, London.

Clarkson, E.N.K. (1998) *Invertebrate Palaeontology and Evolution*. Fourth edition. Blackwell Science, Oxford.

Donovan, S.K. (2017) *Writing for Earth Scientists: 52 Lessons in Academic Publishing*. Wiley-Blackwell, Chichester.

Gould, S.J. (1989) *Wonderful Life: The Burgess Shale and the Nature of History*. W.W. Norton, New York.

Knell, S.J. (2013) *The Great Fossil Enigma: The Search for the Conodont Animal*. Indiana University Press, Bloomington.

Stringer, C. (2011) *The Origin of our Species*. Allen Lane, London.

Tucker, M.E. (2011) *Sedimentary Rocks in the Field: A Practical Guide*. Fourth edition. Wiley-Blackwell, Chichester.

Wyse Jackson, P.N. (2019) *Introducing Palaeontology: A Guide to Ancient Life*. Second edition. Dunedin Academic Press, Edinburgh.

CHAPTER 13

FOSSILS IN CAVES

It is not eccentric to discuss caves shortly after reworked fossils. Like those that are reworked, fossils in caves are, at first glance, out of context. Natural caves are commonly solution features in limestones. Cave fills are invariably much younger than the surrounding rock that hosts the cave. Typical examples in the British Isles are dissolved into Devonian, Mississippian or Jurassic limestones, yet containing a fill of sedimentary rock of Pleistocene age or older. For example, a famous example in the UK is Kirkdale Cave in Yorkshire, investigated as a Pleistocene hyena den in the 1820s; the country rock is Jurassic.

Beware that similar discordances may occur at a smaller scale where a narrow fissure in a rock may have become filled by a younger sedimentary rock. My late colleague, Hal Dixon, brought such an occurrence to my attention. He collected a fossilized free finger of the large Caribbean land crab, *Cardisoma guanhumi* Latreille, from a favourite site in eastern Jamaica, Christmas River. The limestone cliffs here are early Pleistocene, Manchioneal Formation, yet the crab finger was in a lithified red sedimentary rock. We went back and discovered the truth; the finger came from a fissure in the rock and was probably late Pleistocene at the oldest (Donovan & Dixon, 1998).

There are two possible places to find fossils in caves: either within the sedimentary infill, that is, younger than the surrounding rock, or actually within the walls, floor and roof of the cave. An example of each is discussed below – my two favourite caves in Jamaica, both bursting with fossils. The Red Hills Road Cave (RHRC) is an example of the former, with highly fossiliferous, Late Pleistocene siliciclastic sediment infilling a cavity in Miocene limestone. An example of a cave with fossils in abundance in the walls is the wonderfully named Wait-a-Bit Cave in the Eocene of north-central Jamaica.

Red Hills Road Cave

The RHRC, parish of St. Andrew, was only discovered in the 1980s and gives a good demonstration of what gems might be hiding in so many undiscovered caves (Donovan *et al.*, 2013; Donovan, 2017). It is only a minor feature in the highly karstified landscape of Jamaica, but it has yielded the most diverse fauna of Late Pleistocene terrestrial fossils known from the island. The country rock is the White Limestone Group (Miocene) and the cave fill is red, siliciclastic sediment, largely unlithified and with an abundant

Figure 13.1 The Red Hills Road Cave (after Donovan *et al.*, 2013, fig. 3A). This is a general view of the cave (centre), with Professor Chris Paul providing a scale. The remaining shallow section of the cave is shaped like the side of a bottle; it would have extended through what is now the main road (foreground). A small tree grows on the remnants of the sedimentary infill. Image taken during the early or mid-1990s.

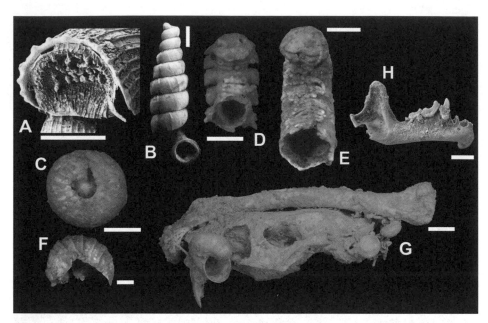

Figure 13.2 Representative fossils from the Red Hills Road Cave, parish of St. Andrew, Jamaica (after Donovan, 2017, fig. 2). (**A, B**) Scanning electron micrographs of land snails; specimens lost. (**A**) *Fadyenia lindsleyana* (**C. B**. Adams), detail of operculum preserved *in situ* in aperture of shell. Scale bar represents 500 μm. (**B**) *Geoscala costulata* (**C. B**. Adams), adult shell in apertural view. Scale bar represents 1 mm. (**C–F**) Millipedes (**C—E**) and isopod. (**C**) *Cyclodesmus* sp. cf. *C. porcellanus* Pocock, RGM 789 611, lateral view of enrolled specimen. (**D**) *Caraibodesmus verrucosus* (Pocock), RGM 789 607, ventral view of the anterior segments, showing the head, and basal attachments for legs and antennae. (**E**) *Rhinocricus* sp., RGM 789 601, ventral view of anterior showing the head, antennae and legs. (**F**) *Venezillo booneae* Van Name, RGM 789 615, lateral view showing the ability to roll into a ball and the slightly raised confluent tubercles on the dorsal surfaces of the segments. Scale bar represents 1 mm. (**G, H**) Vertebrates. (**G**) *Geocapromys brownii* (Fischer), skull of a rodent, cemented to right humerus of the extinct flightless ibis *Xenicibis xympithecus* Olson & Steadman and various land snails. Specimen in the Geological Museum, University of the West Indies, Mona. Scale bar represents 10 mm. (**H**) *Stenoderma rufum* Desmarest, RGM 632 059, right lateral (labial) view of right mandible of a bat. Specimens in the collection of the Naturalis Biodiversity Center, Leiden (prefix RGM) unless stated otherwise. Scale bars represents 2 mm unless stated otherwise.

fauna. Although now truncated by a road, it was either a fissure or a bottle-shaped cave (Fig. 13.1). Land snails are diverse and commonly well preserved. Tetrapods are invariably disarticulated; they were washed in during storms, drowned and rotted away (Donovan, 2017). In contrast, millipedes and isopods are beautifully unspoiled in three dimensions, preserved by calcite that crystallized on their exoskeleton. This is the richest site for terrestrial taxa in the Jamaican fossil record (Fig. 13.2).

Vertebrates, all preserved as disarticulated bones or teeth, include indeterminate amphibians, small lizards, birds and small mammals. Birds include limb bones of the extinct ibis, *Xenicibis xympithecus* Olson & Steadman (Fig. 13.2G), but await monographic treatment.

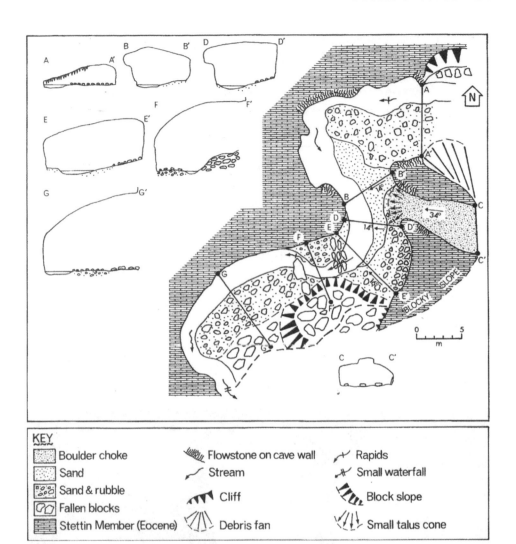

Figure 13.3 Cave survey and selected passage cross-sections (**A–A'** to **G–G'**) of Wait-a-Bit Cave, parish of Trelawny, Jamaica (after Miller & Donovan, 1996, text-fig. 2). The thick dashed line to west of **E'**, and south of **F'** and **G'**, marks the edge of the limestone overhang from the north-west. There is a side passage with an entrance at **C–C'**.

Mammals have been described in some detail, including the Jamaican hutia, *Geocapromys brownii* (Fischer) (Fig. 13.2G), a rodent that is still extant. Fossil bats are moderately well known from the Pleistocene of Jamaica and four species, all rare, have been described from the RHRC (Fig. 13.2H), including one new species.

Jamaica has a highly diverse fauna of extant terrestrial gastropods with over 500 species, with many endemic to limited areas. The Pleistocene terrestrial gastropods from RHRC is rich with 62 species (Fig. 13.2A, B). This is more diverse than the Recent fauna in the area, but the RHRC snails presumably accumulated over a long time, with a low proportion of fossil species known to still live in the immediate vicinity of the cave. One reason for the high diversity of the cave fauna is the occurrence of several species not previously reported from Jamaica.

The macro-arthropods of the RHRC consists of thirteen or more taxa, including millipedes (at least four species) (Fig. 13.2C–E), isopods (four species) (Fig. 13.2F), decapods (one species) and insects (four species). Most of these taxa are classified in open nomenclature, which is a function of the preservation of most of these fossils with a sugary coat of calcite on the external surface. Preservation has favoured those taxa with a small proportion of calcite in their exoskeleton in life, which facilitated inorganic encrustation with calcite post-mortem. The ostracods are the only arthropods from the RHRC that await description.

There is no strong indication that any of the fossil macroinvertebrates known from the RHRC were obligate cave dwellers. They provide information about the surrounding environment in the Late Pleistocene rather than in the cave.

Wait-a-Bit Cave

Wait-a-Bit Cave, parish of Trelawny, Jamaica, has several points of geological and geomorphological interest (Fig. 13.3). It is a through cave, with a stream that flows in

Figure 13.4 The irregular echinoid *Eurhodia matleyi* (Hawkins), Wait-a-Bit Cave, parish of Trelawny, Jamaica (after Miller & Donovan, 1996, pl. 3, figs 1–5); all Natural History Museum, London, specimens (prefix BMNH). This species forms a thin, echinoid-rich horizon. (A, D) BMNH EE5194. (A) Apical view. (D) Lateral view (anterior to right). (B, C, E) BMNH EE5193. (B) Apical view; note growth deformity in anterior petal. (C) Lateral view (anterior to left). (E) Oral view. Specimens coated with ammonium chloride. All scale bars represent 5 mm.

at one end and out the other. The local geology is the Eocene Yellow Limestone Group, Chapelton Formation, which dips towards the end at which the river flows in. Thus, at this end the roof is low and the entrance no more than a big slot. At the outflow end the roof is so high that it forms a cliff overhung by tropical creepers (Miller & Donovan, 1996).

The Eocene palaeontology is mixed in the best way, with a fascinating diversity of macrofossils to be found in the walls of the cave and in the river bedload. Echinoids (Fig. 13.4) and molluscs are particularly common, including the giant gastropod *Campanile trevorjacksoni* Portell & Donovan. A single nautiloid has been described (Donovan *et al.*, 1995), a rare group in the Jamaican Cenozoic. Roger Portell (Florida Museum of Natural History, Gainesville) has found ribs of a primitive dugong (sea cow).

References

Donovan, S.K. (2017) Contrasting patterns of preservation in a Jamaican cave. *Geological Magazine*, **154**: 516–520.

Donovan, S.K., Baalbergen, E., Ouwendijk, M., Paul, C.R.C. & Hoek Ostende, L.W. van den. (2013) Review and prospectus of the Late Pleistocene fauna of the Red Hills Road Cave, Jamaica. *Cave and Karst Science*, **40**: 79–86.

Donovan, S.K. & Dixon, H.L. (1998) A fossil land crab from the late Quaternary of Jamaica (Decapoda, Brachyura, Gecarcinidae). *Crustaceana*, **71**: 824–826.

Donovan, S.K., Portell, R.W., Pickerill, R.K., Robinson, E. & Carter, B.D. (1995) Further Tertiary cephalopods from Jamaica. *Journal of Paleontology*, **69**: 588–590.

Miller, D.J. & Donovan, S.K. (1996) Geomorphology, stratigraphy and palaeontology of Wait-a-Bit Cave, central Jamaica. *Tertiary Research*, **17** (for 1995): 33–49.

CHAPTER 14

BEACHCOMBING

After all my talk of measuring sections and stratigraphic precision, how can I enthuse about such a random sampling method as beachcombing? Before you rip this chapter out and burn it, step back. It is a well-known supposition amongst palaeontologists that the best specimens are commonly found loose, in the float. This is an equivalent observation to the scientific 'law' that when you drop a slice of buttered toast, it commonly falls butter-side-downwards. So, whether you are working in a quarry, a road cut or on the coast, do not dismiss any float specimen out of hand without first examining it closely – it could be the best fossil you will see all day.

Let me give an extreme example of a float specimen leading to greater things. The story may or may not be entirely true (Gould, 1989, pp. 70–75; Yochelson, 2001, p. 49), but it is a good story. Charles Doolittle Walcott, Secretary of the Smithsonian Institution, and a leading expert on Cambrian palaeontology and stratigraphy, was in the field in the Rocky Mountains, and on horseback with his family. Mrs Walcott's horse allegedly stumbled on a float block, derived from higher in the section, and containing abundant and unusual Cambrian fossils. The following field season, searching up-slope, the outcrop of the Middle Cambrian Burgess Shale was discovered.

This is 'the most precious and important of all fossil localities – the Burgess Shale of British Columbia' (Gould, 1989, p. 13) for many reasons, not least its fidelity of preservation (both mineralized and unmineralized fossil plants and animals) and its age, this site recording as it does the first great diversification of multicellular life near the start of the Phanerozoic.

This example is not beachcombing, I freely admit, but does demonstrate the potential importance of some samples collected *ex situ*. Other float fossils will be less significant, but can nevertheless by instructive, scientifically relevant and of aesthetic value. Taking a well-known European example, consider the palaeontological wealth of the Lower Jurassic (Lias) on the Dorset coast around Lyme Regis. Many, many people beachcomb this coastline for fossils, and yet still new specimens appear. In part this is due to the instability of the cliffs, particularly following the winter, when freeze–thaw cycles weaken their integrity. This weakening results in landslides and mudslides, which introduce new rock to the beach. Fine-grained sedimentary rocks are broken down and dispersed by wave action, particularly during major storms, and the larger, heavier fossils are washed

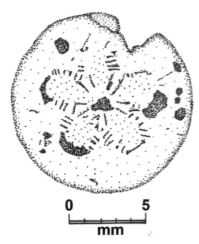

0   5
mm

Figure 14.1 Camera lucida drawing (see Chapter 29) of a columnal of the Upper Cretaceous crinoid *Austinocrinus bicoronatus* (von Hagenow) collected from the beach between Overstrand and Cromer, north Norfolk (after Donovan, 2012, fig. 2). Collection of the Naturalis Biodiversity Center, Leiden, number RGM.621009.

out and concentrated. The beachcombing palaeontologist with the sharpest eye will spot the best finds.

But this brings us to an obvious consideration: that is, how do you spot a tiny fossil by your feet while you are standing up? Large specimens can be found while standing, such as large Chalk sea urchins, but it is a stance that may limit your collecting to the big and obvious. I did once collect a crinoid columnal the size of a shirt button while standing up and looking down (Fig. 14.1), but this was an exception. The answer practice, practice, practice is obvious and too glib, but it is also true. The first lesson is, do not stand up. I like my face about 20 cm from a rock surface when I am searching for fossils, but that is not always possible on a beach. However, whenever the beach surface is promising for recumbent collecting, be happy crawling.

Look for local concentrations of likely-looking pebbles and rock fragments, and sit down on the edge of it. Pick up the beach clasts and examine them with the naked eye and a hand lens. Discarded specimens can be piled as a small cairn, away from the specimens you are examining, and which you can ignore subsequently.

The trick of beachcombing for the palaeontologist is in recognizing that your carefully selected beach has the potential to be a concentration point for fossils derived from many directions. There will be fossils derived (assuming these are fossiliferous rocks) from the cliffs behind the beach, from offshore (rocks exposed below the sea and, in consequence, not easily accessible to you), and from laterally, transported by longshore drift. Gravity and waves are thus your friends, bringing material to your collection base. For example, in Noord Holland, near where I live, the beach is backed by sand dunes and there is not an *in situ* rock in sight. However, after a major storm, Quaternary mudrocks and peats may be carried onshore. The palaeontological interest of these is the included plant remains, but my personal curiosity focuses on the Recent borings. These may be well preserved and provide comparative material for my interest in ancient trace fossils. For example, slabs of mudrock are commonly pierced by bivalve borings, the ichnogenus *Gastrochaenolites* Leymerie (Fig. 14.2).

You must indulge me, but I am going to introduce two of my favourite beachcombing strands, and which I discuss in more detail later in the book. This is partly to introduce two field areas that may be near you, but mainly to give the reader a feel for what might be found on a good beach. The north Norfolk coast between Overstrand

Figure 14.2 Pleistocene peat pebble with Recent bivalve boring *Gastrochaenolites lapidicus* Kelly & Bromley, RGM.791149, from the strandline of the beach north-northeast of Zandvoort, Nord Holland, the Netherlands (after Donovan, 2013, fig. 2A). Uncoated specimen, boring upper centre. Scale in cm.

and Cromer has been one of my happy hunting grounds for many years (Chapter 47). Chalk fossils are uncommon, but varied; the more common taxa include echinoids, belemnites and sponges. I am also interested in the Recent borings along this coast, both in limestone clasts and in fossils. I was 'hooked' by my first find, just a broken piece of belemnite in a Chalk cobble. The belemnite guard had been bored post-mortem while it lay on the Late Cretaceous sea floor, infested by clionaid sponges, architects of the complex boring *Entobia* isp. These borings were filled by lithified Chalk. Over 65 million years later, the fossil in a loose Chalk cobble was lying on the modern sea floor when it was bored by a clionaid sponge. Modern *Entobia* borings crosscut ancient chalk fills of *Entobia* (Fig. 47.2A, B).

Cleveleys on the Fylde coast in north-west Lancashire (Chapter 48) is an exceptional collecting ground for any geologist, not just the palaeontologist (Ellis, 1968, p. 144; Fig. 48.2 herein). During the Pleistocene, glaciers ground their way south from the Lake District, transporting ripped-up clasts of diverse rock types. They were sealed in the local 'boulder clay' when the glaciers retreated and are now being reworked by coastal erosion. Clasts, all well-rounded by wave action, litter the beach in abundance. One of the commoner lithologies are Lower Carboniferous (Mississippian) limestones. In some of these there are Recent borings to divert my interest, but there are also Mississippian fossils, most notably several handsome species of corals (Fig. 48.4).

References

Donovan, S.K. (2012) Taphonomy and significance of rare Chalk (Late Cretaceous) echinoderms preserved as beach clasts, north Norfolk, UK. *Proceedings of the Yorkshire Geological Society*, **59**: 109–113.

Donovan, S.K. (2013) A distinctive bioglyph and its producer: Recent *Gastrochaenolites* Leymerie in a peat pebble, North Sea coast of the Netherlands. *Ichnos*, **20**: 109–111.

Donovan, S.K. & Lewis, D.N. (2010) Notes on a Chalk pebble from Overstrand: ancient and modern sponge borings meet on a Norfolk beach. *Bulletin of the Geological Society of Norfolk*, **59** (for 2009): 3–9.

Ellis, C. (1968) [first published 1954] *The Pebbles on the Beach*. Faber and Faber, London.

Gould, S.J. (1989) *Wonderful Life: The Burgess Shale and the Nature of History*. W.W. Norton, New York.

Yochelson, E.L. (2001) *Smithsonian Institution Secretary, Charles Doolittle Walcott*. Kent State University Press, Kent, Ohio.

CHAPTER 15

COMMON SENSE IN THE FIELD

This is an intentionally short chapter. The original plan was to call it 'Health and safety', but I thought this would put off at least some readers. Health and safety has a bad name, because many equate it to increased paperwork that states the obvious, an additional hurdle to keep you from doing something you regard as interesting, important and, dare I say it, just plain fun. One colleague at a leading museum would shy away from the excessive paperwork necessary just to be granted a day in the field by taking eight hours' annual leave, instead, and having a 'holiday' in a quarry. I can only sympathize with this position, as an equal amount of paperwork was involved to take a day trip to a chalk pit in the North Downs or a month in the jungles of south-east Asia.

What I want to emphasize is that much of what sits under the umbrella of health and safety can equally well be regarded as applied common sense. Accidents and slips may occur, but let's keep them to a minimum, both in number and severity. Health and safety should be a state of mind, an awareness of your environment.

Using tools in the field

Consider hammers and chisels. Never bang two geological hammers together, using the wedge-shaped end of one as a chisel. The heads of geological hammers are equally hard, formed of hardened steel. Striking them together may cause them to chip or shatter, producing flying splinters of steel that can damage an eye. If you have fine work to cut a fossil out of a rock, use a hammer and cold chisel. A cold chisel is made of tempered steel that is softer than a hammer head and is more likely to deform than shatter. This is why well-used chisels have a burred head. A cold chisel is also a more subtle tool than just hammering, permitting a more precise excavation.

I have not emphasized the use of a hammer, as much collecting can be done without it and, in truth, many people make a rock face 'ugly', hammering to little purpose and breaking the good specimens that they do find. But I realize that hammering is a necessary part of geology. So, be sensible. Wear some sort of glasses or goggles as protection to save your eyes from flying chips. Safe hammering develops with practice. Do not hold the shaft too rigidly or you will soon be tired, the vibrations on impact jarring your arm. Heavy, imprecise blows will rarely liberate that fine specimen.

One of the advantages of being more of a beachcomber, like me, is that hammering is kept to a minimum. Even if I do need to hammer, I must wear bifocal glasses with plastic lenses just to see – not quite as worthy as protective goggles, but still a good backup.

Various codes of conduct for the field palaeontologist are available on-line. For over 40 years I have referred to the Geologists' Association Code for Fieldwork. This is available at the Association's website under Publications (www.geologistsassociation.org. uk), along with a Foreshore Code of Conduct. There are other related codes, such as the Scottish Fossil Code (www.nature.scot) published on-line by Scottish National Heritage.

In case of emergency

Do not forget the emergency supplies. You ought to make sure that these are adequate, but minimal, to keep the weight and bulk down. A bottle of water is always useful for drinking or washing wounds (or specimens). A small first-aid kit, such as some sticking plasters and a tube of Germolene antiseptic cream, or similar, will be needed sooner or later. Some long-lasting snacks are desirable, such as cereal bars, chocolate and/or a bag of raisins, the last a no-nonsense way to take fruit into the field. You can leave all of these in your backpack for months and they will still be edible when eventually needed. A diabetic such as your author keeps a tube of Dextrose tablets in his backpack, as well as my medication. I also keep painkillers, such as Paracetamol, and Rennies to hand. Sun cream and insect repellent are light to carry, and may be essential. I normally wear a hat with a broad brim and at least carry some gloves. The hat keeps the sun out of my eyes, the rain off my glasses and my head warm.

Do you know where you are? Do not get lost. Have a good topographic map of your field area (1:50,000 scale at least), a GPS that you know how to use and a field guide, if available. Maps have the advantage over a GPS in that there are no batteries to go flat, but you should also have spare batteries. Similarly, make sure your mobile phone has an adequate charge.

But do not fill your backpack with 'clutter' before you even get into the field. The object of palaeontological fieldwork should be to accumulate data about fossils, including collecting specimens. Fossils are rocks, and may be heavy and bulky. But big specimens can be collected nonetheless:

A notable field excursion visited Chislet Colliery in the Kent Coalfield, on 28th September, 1958, where collecting from the spoil tips was successful ... The meeting was arranged on behalf of Colin Brooker, a young lad keenly interested in palaeobotany. Nine members attended and were soon passing specimens to Colin for his approval. His rucksack became bulkier and bulkier. Towards the time for departure, [Joe Collins] saw him a little way off staring at an object at his feet. Curious, Joe wandered over. 'Oh, I'd dearly love that' said Colin, pointing to a lump of *Sigillaria* trunk about a third the size of a cement sack, with an additional spur of rock the length of one side. Joe pondered:

'Hmm, if we can get that spur off and carry the trunk to the bus stop, would that suit you?' Colin needed no second suggestion and, after a couple of whacks with a hammer, the spur came cleanly away. A schoolboy member, Ted Rose (later Dr E.P.F. Rose of the University of London), was nearby and, calling him over, he readily complied with the plan. The trunk was swathed in newspaper to camouflage it and disguise its obvious considerable weight. Other members took charge of Colin's bulging rucksack. The way out from the exposure was via a high iron gantry and, even before they reached it, Ted and Joe were counting their steps with a regular cry of 'Over to you'; so it continued to the bus stop. As the bus drew up the parcel was passed to a totally refreshed Colin, who stepped lightly(!) aboard; others swore the platform tilted. Little is known of events to Bromley South station, but the last sight of Colin was of him staggering along the platform, counterbalanced front and back. On reaching home, the first thing his mother said was 'You're not having that in the house!' (after Donovan & Collins, 2020, p. 58).

References

Donovan, S.K. & Collins, J.S.H. (2020) In the field with Joe: early excursions of the Freelance Geological Society. *Geology Today*, **36**: 53–58.

CHAPTER 16

COLLECTING WITH A CAMERA

'excavation without publication is simply destruction'
(Professor Graham Clark quoted in Pryor, 2019, p. 14).

I like the way that archaeologists think. If you excavate, then record and publish. But the nature of the archaeological record is different from the ancient fossil record. An archaeological site is of limited spatial extent, whereas a sedimentary bed preserving a common fossil, X, may extend over a broad area, tens or hundreds of square kilometres. So, if I collect a specimen of X, who cares? There should be plenty to go around.

Well, perhaps we should all care. The older I get, the fewer specimens I collect. There are many reasons for this, not least storage. I worked for various natural history museums, so I had more than adequate shelf space readily available. Yet it is always easier to collect specimens than to deal with them – clean them, separate them out into species groups, write labels with all relevant data, describe and publish them if they are new species or exceptional specimens. The list goes on and on. It is much easier not to collect.

This attitude may appear heretical to some, but the world's natural history museums are stuffed to the gunwales with specimens people considered 'important', but which, once installed in a drawer, were left to accumulate dust. Out of sight, out of mind – so why are they important? Remember, these museums are nature's treasure houses, but is this treasure being adequately treated if all that they do is hoard?

What of the collections made by amateurs? Conscientious amateurs, such as my late friends Stanley Westhead and Joe Collins (Donovan, 2012; Donovan & Mellish, 2020), had well-curated, well-labelled collections that they bequeathed to the Natural History Museum in London. Yet how many collectors lose interest and change direction? How many collections are destined for the skip rather than the local or national museum?

Another good reason for not collecting more than is essential is environmental degradation. In short, the more a site gets hammered, the uglier it looks. Anyone who has seen the type locality of the Ludlow bone bed will know what I mean; it has been reduced to a deep slot in the rock face. There are many tales of a significant specimen left in place for teaching, for student visits, for everyone's pleasure, until it is gone, or worse, damaged beyond repair by some ass using a hammer while their brain is in neutral.

OK, tirade over. It took me three lunchtimes to write the above, so I was not just having a bad day. If I am encouraging you not to collect in excess, what are the alternatives? As you will deduce from the title of this chapter, I advocate the advantages of photography in the field.

Today, it is cheap. I wrote my Ph.D. in the age of wet photography. I had to buy chemicals and photographic paper in bulk, and spent days and weeks in the darkroom. With digital photography there is no film to process, no days in the darkroom and no necessity to make multiple prints at different exposure times in order to produce one print that is just right. What used to take days now takes hours, at most. With Photoshop you can manipulate all your images to be just right with a fraction of the time and effort that I had to expend in the early 1980s (Figs 16.1, 16.2).

Photos also offer advantages of storage. Fossils need to be stored in three dimensions, in a drawer or cabinet. They are rocks, after all. Yet digital photographs are stored in your computer. Your digital images will require little more file space to store them if there are one or a thousand images, whereas a thousand ammonites will need a considerable storage cabinet. Further, the documentation of your digital files is as important as for your specimens – be organized.

Obviously, not everything can be photographed, and you will want to collect those specimens in which you have a particular interest. Say, for example, your main interest in palaeontology is Mesozoic cephalopods – ammonites, nautiloids and belemnites. You will want to collect prize specimens of all these groups (and, perhaps, photograph them at home; see Chapter 28), but what of the associated gastropods, bryozoans and crinoid columnals? Would good, in-focus images of these fossils taken in the field be adequate?

A further advantage of photographic images is that they are easy to exchange. A week rarely goes by without an email arriving that boils down to either, 'Steve, I found this specimen in Anonymous quarry and have little idea of what it is. Do you have any insights?' or, 'Steve, this is obviously a crinoid [my favourite group], but I suspect it is a new or rare species. Would you like to describe it with me?' Both types of email are exciting to receive.

Yet another part of collecting by camera is that you are already part of the way to a publishable paper or image for your website. Take good field photographs, perhaps supplemented by images taken at home of specimens that you collected and relevant diagrams such as a locality map (Chapter 29), and you are halfway to a publishable article: all you need are the words (Chapters 31, 35, 36).

References

Donovan, S.K. (2012) Stanley Westhead and the Lower Carboniferous (Mississippian) crinoids of the Clitheroe area, Lancashire. *Proceedings of the Yorkshire Geological Society*, **59**: 15–20.

Figure 16.1 Not palaeontological, but certainly geological and too big to take home (even if the zoo would let you). A mock inselberg in the African vulture enclosure, Blijdorp Diergaarde, Rotterdam, the Netherlands (after Donovan, 2019a, fig. 1). A vulture nesting box (upper left), poles supporting the netting of the aviary, trees and the overhead electrification of the main Rotterdam to Utrecht railway line are all apparent in the background. Birds for scale (left of centre) are an African secretary bird and a Dutch crow. An exposure of this size is uncollectable except with a camera.

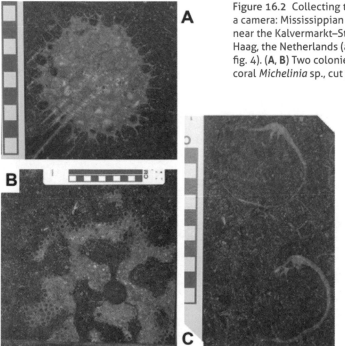

Figure 16.2 Collecting the uncollectable with a camera: Mississippian fossils in paving slabs near the Kalvermarkt–Stadhuis tram stop, Den Haag, the Netherlands (after Donovan, 2019b, fig. 4). (A, B) Two colonies of the tabulate coral *Michelinia* sp., cut parallel to their bases. (C) Two large, disarticulated brachiopod valves preserved in the same orientation with the umbo towards the top of the page. Scales in cm.

Donovan, S.K. (2019a) Urban geology: An inselberg in Rotterdam. *Deposits*, **57**: 12–13.

Donovan, S.K. (2019b) Urban geology: Mississippian in the Mainstreet. *Geology Today*, **35**: 135–139.

Donovan, S.K. & Mellish, C.J.T. (2020). Mr Joseph Stephen Henry (Joe) Collins, 1927–2019. *Bulletin of the Mizunami Fossil Museum*, **46**: 103–114.

Pryor, F. (2019) *Paths to the Past: Encounters with Britain's Hidden Landscapes*. Penguin, London.

CHAPTER 17

BUYING SPECIMENS

Fossils for sale are part of the retail twenty-first century. Most museum gift shops sell geological specimens to their visitors, some of which may actually be of scientific value (Donovan & Lewis, 2004). Rock and fossil shops are specialist outlets that may be found in areas of noteworthy geology and elsewhere, and online. Specimens may be local or exotic, imported from overseas. Details of locality and horizon provided with these specimens are commonly minimal, but rarely erroneous.

I include this subject here because there is always a temptation to buy a thing of beauty. If trilobites are your heart's desire, then a fine specimen in the window of a rock and fossil shop may be too much for you and your bank account to resist. I cannot be critical – I have succumbed to temptation more than once. But beware—many countries, such as China and Brazil, have strong legislation in place to control the illegal sale of their fossil heritage overseas, but it still happens, particularly their fossil vertebrates. Ensure that the specimen you are buying has been collected ethically.

You should not let puritanical museum curators influence you, either. I remember one article by a curator and palaeontologist who was entirely critical of the professional collectors who sold fossils for a living rather than donating them to a museum. Some years later, after moving to a national museum, the same curator was happy to fly to the USA and buy specimens from said professional collectors for display. New job, new morality.

I regard buying fossils as something we are all likely to do, palaeontologists or not, to grace our shelves or drawers or cabinets with fine items that connect us with the Earth's distant past. We all want to collect – it is in the blood of all palaeontologists – but sometimes that is just not possible. I taught at the University of the West Indies for over twelve years and introduced fossil arthropods to my students through everyone's favourite group, the trilobites. The Jamaican rock record is Cretaceous to Cenozoic, so there are no trilobites. Every two to three years I would go to the annual meeting of the Geological Society of America, in Denver and elsewhere. I would always let my second-year class know that trilobites (calymenids from Morocco) would be on sale for about US$3 – who wants one? I would be coming back from the USA with at least a dozen 'orders' of trilobites.

Of course, things sometimes cut the other way. Before we met, my partner, Karen, had a taste for attractive crystals that she bought to decorate her apartment. Now she saves her money and waits for me to discover some attractive rock or mineral as an aside from my palaeontological fieldwork.

But beware. There are both scrupulous and unscrupulous sellers of fossils. At least some of the trilobites that I bought for my students were obviously glued together from two or more incomplete specimens. This presumably happened in Morocco. Nobody seemed to mind. Closer to home, I found some West Indian fossil echinoids pretending to be from the Chalk in a rock and fossil shop.

Case study: polished echinoids

These notes are adapted from Donovan (2014). In July 2014 I visited a rock and fossil shop on the Isle of Wight, run by a friend, in search of research specimens. Among the intriguing specimens on this visit was a small collection of polished irregular echinoids, labelled as coming from the Chalk. This provenance was not unexpected for the island, but the specimens were immediately recognizable as being in disagreement with this information. Two specimens (Fig. 17.1) were bought for further investigation because their age and provenance appeared debatable. These specimens are now registered in the collections of the Naturalis Biodiversity Center, Leiden, the Netherlands (prefix RGM).

Both specimens were identified as Cenozoic, probably from the Caribbean or south-east USA. Presumably, neither was well preserved, otherwise they would not have been polished. Polishing has not improved their systematic identity and neither is confidently identifiable below the level of genus.

The test, RGM.791773, is the most revealing, belonging to *Oligopygus* sp. (Fig. 17.1A, B), and a genus of limited stratigraphic and geographic distribution. Although polishing has obliterated surface detail, such as most of the tuberculation, several diagnostic features were still identifiable (Smith & Kroh, 2011). The oligopygids are an extinct sister group to the clypeasteroids, such as the sand dollars, that are limited geographically to the Caribbean, south-east USA and Peru. They are similarly stratigraphically limited and are only known from the Eocene, going extinct during the Eocene–Oligocene extinction events. No British Cretaceous echinoids are close in morphology to RGM.791773.

The second specimen, RGM.791774, is presumed to be Eocene, too, although it represents a genus with a range Eocene to Recent. *Eupatagus* sp. (Fig. 17.1C, D) is a spatangoid (heart urchin) genus with a global distribution at low to mid-latitudes, but is particularly common in the Eocene of the Caribbean and southern USA. The posterior is broken and the periproct was situated in this region. Again, despite the loss of surface detail due to polishing, diagnostic features are identifiable (Smith & Kroh, 2011). Comparison with the British Cretaceous echinoids revealed no close morphological similarity to any spatangoid taxon.

These specimens are demonstrably not from the Upper Cretaceous chalk. They bear no close comparison with any of the many irregular echinoids known from the Cretaceous of the British Isles and are undoubtedly Cenozoic. One specimen is undoubtedly Eocene (*Oligopygus* sp.), the other probably so. If this is the case, they are most probably from the

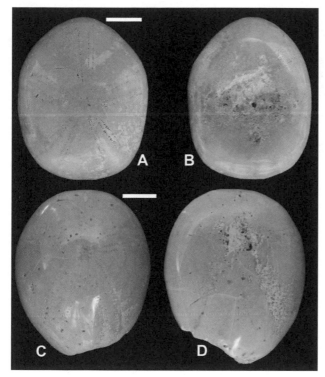

Figure 17.1 Polished echinoids from the Paleogene of the Antilles or south-east USA, not from the Chalk of the Isle of Wight (after Donovan, 2014, figs 1A, B, 2A, B). (**A**, **B**) *Oligopygus* sp., RGM.791773. (**A**) Apical view. (**B**) Oral view. (**C**, **D**). *Eupatagus* sp., RGM.791774. (**C**) Apical view. (**D**) Oral view. Specimens uncoated. Scale bars represent 10 mm.

southeastern USA, where collecting, trading and selling fossils is more common than in the Antilles. That the owner of the shop, a vertebrate palaeontologist, did not recognize this discrepancy is not surprising; it may be that the polisher was similarly misled. Even though the stratigraphic information was faulty and the implied locality (Isle of Wight) out by several thousand kilometres, anyone purchasing one of these echinoids would have gained an exotic specimen for their collection, albeit unknowingly.

The relevance of this tale to collectors is obvious. Specimens that are being bought for display need to have the best possible evidence of provenance. Reference to a recognized expert would be an extra check that the associated documentation of locality and horizon is at least plausible.

References

Donovan, S.K. (2014) Chalk or Eocene echinoids? Implications of a chance observation. *Geological Curator*, **10**: 67–70.

Donovan, S.K. & Lewis, D.N. (2004) Palaeoecology in the museum gift shop. *Proceedings of the Geologists' Association*, **115**: 367–370.

Smith, A.B. & Kroh, A. (eds) (2011) *The Echinoid Directory*. World Wide Web electronic publication. http://www.nhm.ac.uk/research-curation/projects/echinoid-directory.

SOME THEORETICAL
ASPECTS

CHAPTER 18

PALAEOECOLOGY 1: THE ORGANISM

The objectives of palaeoecology

Ecology is 'the interrelationships between organisms and their environment and each other; the study of these interrelationships' (Lawrence, 1989, p. 153). Thus, it is the study of interactions between organisms, and their physical and chemical environment. An ecologist studies these interactions at the present day, observing relationships between organisms, and measuring physical and chemical parameters such as temperature, water salinity, etc. The part of the environment that is being studied by an ecologist is called an ecosystem and includes all of the associated organisms, plus the inorganic and physical components.

In modern ecosystems, there are numerous relevant measurements that may be made to define the biological, chemical and physical parameters that mark its extent and limits. We must recognize that many of these can only be inferred, not directly measured, when considering fossil organisms. So, you might ask, is it possible to make a worthwhile contribution by attempting to interpret the ecology of fossils? Yes, providing shortcomings are recognized, studies are selective and the right questions are asked; I emphasize this last point. The quest for the ecology of fossils is called palaeoecology, which 'studies the relationships of fossil organisms to past physical and biological environments' (Brenchley & Harper, 1998, p.1). It is divided, simply, into two interrelated areas of study:

- Palaeoautecology is the study of the ecology of individual fossil organisms, especially functional morphology.
- Palaeosynecology is the study of communities of the past and their relationships, both to the environment and themselves, where a community is a natural grouping of two or more species that live in close association.

Palaeoecology must always be doomed to at least partial failure. Much of the information that can easily be determined by the modern ecologist is lost in the rock record. The principal limitations have been summarized by Ager (1963, pp. 10, 12). Nevertheless, it is still possible to make valid palaeoecological deductions, both of individual organisms and communities, based on reasoned arguments.

Palaeoautecology

Palaeoautecology is often involved in interpreting the mode of life of a fossil by comparison with modern related or functionally similar organisms. Nobody has ever seen a Mesozoic winged reptile fly, yet comparison with modern flying vertebrates such as birds and bats leaves us in little doubt that a pterosaur was capable of gliding and/or flapping flight. The 300+ species of extant brachiopod are all marine, a habit indicated for the 30,000+ fossil species by various lines of evidence.

Ancient organisms belonging to extant groups may nevertheless be so different that comparison can only be made in a broad sense. For example, asteroids (starfishes or sea stars) have changed little in appearance since they first appeared in the early Ordovician and until recently it was assumed that all members of the group were functionally similar. It has now been proposed that Palaeozoic asteroids were deficient in features typical of living members of this group such as muscular arms, an eversible stomach and a flexible mouth frame. If a modern asteroid is turned upside down, it is able to right itself using its muscular arms and tube feet.

Palaeozoic asteroids may have lacked such a 'righting' mechanism. The asteroids remained a minor group during the Palaeozoic, but underwent a major radiation in the Mesozoic, after the evolution of advanced features.

Functional morphology

Functional morphology is the study of how whole organisms and their included structures are, and were, adapted to fulfil their ecological requirements. In a living organism the function of an organ or structure can be determined by direct observation. For example, the legs in humans are well suited to support the weight of the body and for locomotion. Legs and feet are used during other, perhaps less vital tasks, such as kicking a football and swimming. It is therefore apparent that structures within organisms can have multiple functions and that functions can be ranked in order of importance.

Functional morphology of fossil organisms can often be determined by comparison with living analogues. The methodologies used may be placed into two broad, inferential categories that I shall call uniformitarian and 'non-uniformitarian'. For example, consider the function of the legs in fossil hominids. By comparison with modern humans and other primates it is certain that hominid legs served the functions of supporting the body and locomotion. This is a basic, uniformitarian inference. What must also be considered is whether the hominid walked upright like modern humans, whether it stooped forward or perhaps even required the hands for support in a 'four-legged' gait, or were obligate climbers. The latter options are not known in modern man, but are found in

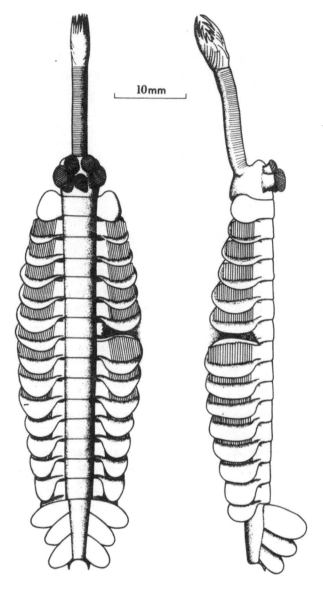

Figure 18.1 *Opabinia regalis* Walcott (after Whittington, 1980, fig. 10). A reconstruction in dorsal (left) and lateral views. Eye lobes are cross-hatched. A lateral lobe and gill have been removed to show succeeding lobe and gill.

10mm

other anthropoids. Deductions of this sort can be made on the basis of features such as arm length, spinal curvature and head angle, and are still uniformitarian, although it is necessary to cast the net wider to find suitable functional analogues.

Now consider the functional morphology of an organism with no living relatives. *Opabinia* Walcott (Figure 18.1) is an organism from the Middle Cambrian that bears a tentative resemblance to an arthropod, but lacks jointed appendages. However, its segmented body suggests that it may belong to some extinct group related to arthropods and/or annelids. How can the functional morphology of the various structures on this organism be deciphered? The body is divided into three regions – a head, a thorax and a tail. The body is bilaterally symmetrical.

Animals with bilateral symmetry have the sensory organs concentrated anteriorly. The bulbous head of *Opabinia* bears five mushroom-like protuberances that have a lattice-like structure similar to the compound eyes of arthropods. These are thus reasonably interpreted as eyes situated on the upper (= dorsal) surface. Anterior to the head is an annulated, probably flexible tube, similar in appearance to a vacuum-cleaner hose, bearing a pincer-like termination that suggests a feeding function. Bilaterally symmetrical organisms are usually vagile, not sessile, and the flexible, jawed feeding tube of *Opabinia* implies that it may have been an active predator or, at least, scavenger. *Opabinia* had a lobed thorax, with the rear three lobes raised up as a tail, which may have been adapted for swimming. The annulated thorax was undoubtedly flexible and any thrust gained during locomotion may have been aided by the lobes. Parallel structures on the lobe surfaces may have been gills.

Because it lacks any living relatives, the interpretation of *Opabinia* is 'non-uniformitarian' of necessity. It could not be compared directly with any living organism. However, in interpreting individual features of this animal, comparison was made with modern, unrelated organisms and even man-made structures (vacuum-cleaner hose) to determine the probable function of separate parts and thus build up a picture of the possible palaeoautecology of the organism. Essentially, a number of uniformitarian parts were used to construct a 'non-uniformitarian' whole. This methodology would have to be altered to encompass more speculative diagnoses if the fossil organism included structures for which we could recognize no modern analogue. Note that if a particular shape or structure is well suited to a given function, it is likely to be well suited to the same task in an unrelated taxon. Thus, fast-swimming vertebrates such as sharks, dolphins, Mesozoic ichthyosaurs and the Cretaceous bird *Hesperornis* all have or had a similar streamlined shape (Ager, 1963, fig. 1.1). Ostracods, bivalves and brachiopods all gain an advantage from having a hard, protective shell, although functionally these organisms are somewhat different.

References

Ager, D.V. (1963) *Principles of Paleoecology*. McGraw-Hill, New York.

Brenchley, P.J. & Harper, D.A.T. (1998) *Palaeoecology: Ecosystems, Environments and Evolution*. Chapman & Hall, London.

Lawrence, E. (1989) *Henderson's Dictionary of Biological Terms*. Tenth edition. Longman Scientific & Technical, Harlow, England.

Whittington, H.B. (1980) The significance of the fauna of the Burgess Shale, Middle Cambrian, British Columbia. *Proceedings of the Geologists' Association*, **91**: 127–148.

CHAPTER 19

PALAEOECOLOGY 2: ORGANISM MEETS ORGANISM

Much of the scope of palaeosynecology (see Chapter 18) involves the comparison of living and fossil communities. Groups of fossils that are preserved together in close association, either in one bed or upon a bedding plane, are called assemblages. Assemblages represent groupings of contemporaneous organisms, unless there has been contamination by reworking (see Chapter 9). Palaeosynecology also considers relationships between organisms that are intimately related, such as a parasite and its host. The ecosystem may thus be larger or smaller, depending upon the questions being asked.

The analysis of any community is a bedding plane investigation, but assemblages are not necessarily equivalent to ancient communities, because post-mortem processes often lead to the transport of shells and other skeletal fragments away from their original life position. For example, consider the range of marine shells that are found washed up on a beach, an environment essentially inhospitable to them. Assemblages composed of fossils removed from their original living position (= allochthonous) are called death assemblages, that is, groupings of remains that have become associated due to post-mortem processes (see below). In contrast, a life assemblage includes an association of organisms that lived together during life and have been preserved in such a way that relationships between fossils are more or less retained (= autochthonous). A life assemblage is equivalent to a fossil community; a death assemblage is not, generally comprising organisms from more than one community, or perhaps from one community that has been transported to a facies different from that in which it lived. A mixed assemblage is generally considered to be a life assemblage that has been contaminated with allochthonous components.

Death assemblages

Death assemblages are produced by processes of physical transport and biological reworking. Physical transport can vary from the small scale, such as shells being rolled along by unidirectional bottom currents, to large scale, such as storm reworking and transport of shallow-water shelf organisms offshore, or the transport of terrestrial animals and plants into the marine environment during a flash flood. For example, the only known stratigraphic unit for terrestrial plants and land snails in the Pliocene of Jamaica is the Bowden Member, Layton Formation, which was deposited in 100+ m water depth (Goodfriend, 1993; Pickerill *et al.*, 1998; Locatelli *et al.*, 2018).

Biological reworking is generally local and is due to bioturbation, or mixing, of sediments produced by burrowing organisms. However, as an example of large-scale biological reworking, concentrations of hippopotamus bones of Ipswichian (= last interglacial) age in upland caves in Britain are thought to have been accumulated away from their normal river environment due to the action of scavenging hyaenas (Rudwick, 2014, pp. 122–125). In general, physical transport tends to mix organic debris laterally, that is, in space; bioturbation tends to mix fossils vertically, or in 'time'.

Death assemblages may be recognized by the presence of disharmonic associations of fossils (for example, terrestrial and marine organisms occurring together) and/or by the presence of disarticulated or abraded fossils. The latter criterion must be used with care, because some organisms disarticulate very rapidly on death, such as echinoderms (Donovan, 1991), so that a slowly buried life assemblage may bear a superficial resemblance to a death assemblage.

A death assemblage: the Paleogene London Clay Biota.

While death assemblages appear to be of limited use to the palaeoecologist, they have the feature that they 'average' local environments, so that a large diversity of organisms from a broad area is concentrated in a single deposit. Further, the range of fossils present may give a broad indication of the local palaeogeography. Thus, a lake samples the area enclosed by the rivers that feed it. Similarly, a marine deposit that includes abundant terrestrial remains is likely to be near-shore, with a declining proportion of terrestrial organisms as sampling moves away from the ancient shoreline. One of the best known associations of fossil terrestrial and marine biotas is the Paleocene to Eocene London Clay Formation of northern Europe (Davis & Elliott, 1957). The elements of the biota include:

- Marine vertebrates – numerous fishes, marine turtles, sea snakes.
- Marine invertebrates – bivalves, gastropods, brachiopods, bryozoans, ostracods, foraminifers, crabs, echinoderms, solitary corals, nautiloids.
- Terrestrial invertebrates – insects.
- Terrestrial vertebrates – tortoises, crocodiles, birds, mammals.
- Terrestrial flora – over 500 species, mainly recognized from fruits.

The terrestrial flora strongly indicates a tropical environment for the forest from which it was derived. All of the London Clay plant species belong to extant genera, and the general mixture of taxa has been compared with that known at the present day in south-east

Asia. Because of the abundance and general good preservation of terrestrial organisms in the London Clay Formation of south-east England, it is deduced that these were derived from a nearby landmass, to the north-west, and carried out to sea by streams and rivers. While the terrestrial plant remains give the best evidence that Eocene Britain was tropical, various animals, such as nautiloids, crocodiles, turtles, hippopotamus-like mammals, sharks and rays, are also reminiscent of the modern tropics.

Life assemblages

A life assemblage represents a palaeocommunity, with organisms preserved *in situ* or after only minimal transportation. If sedimentation is slow, then even within a life assemblage many shelly organisms may have time to disarticulate after death. However, if sedimentation itself is the cause of death, preservation of even delicate structures may be exquisite. Examples of such rapid burial events include ash falls or turbidity currents. Once buried by even a few tens of millimetres of sediment, many organisms will be unable to dig themselves free and will be preserved where they are entombed. Alternately, local conditions may change to both kill and preserve the biota until burial occurs. For example, oceanic anoxia would have a poisoning effect, killing a benthic fauna, while inhibiting bacterial decay and discouraging scavengers.

A life assemblage: Oyster bed, Round Hill Beds, Jamaica.

The sequence illustrated in Figure 19.1 forms part of the exposure of the Neogene Coastal Group at Round Hill (see also Figs 44.1, 44.2). Bed 6 is 3.3 m thick and dominantly composed of the oyster *Crassostrea virginica*, (Gmelin), preserved variously as broken shell fragments, dissociated valves, recumbent, articulated valves, and upright, articulated valves. Barnacles and juvenile oysters encrust the valves on both the inner and the outer surfaces. Young oysters are particularly prominent on some of the largest, upright, mature specimens of *C. virginica* near the top of the bed. Some shells of *C. virginica* have been bored, probably post-mortem, by bivalves and clionaid sponges. Shells in life position occur throughout this unit, but are concentrated at particular horizons, especially towards the top, where shells reach 400 mm in height. Such shells are amongst the largest *C. virginica* known.

Various lines of evidence show this bed to be a life assemblage. The oyster shells are commonly still articulated and preserved in life position. Encrusting organisms are still in life position, growing on oyster shells. These would have been abraded and scraped off if transport had occurred. Disarticulated oyster shells are probably the result of death by natural causes and subsequent rotting of soft tissues with minimal local reworking.

An unusual point about this assemblage is the dominance of one species, *C. virginica*. This is indicative that environmental conditions were particularly favourable for this taxon, almost to the complete exclusion of all other shelly organisms that are as easily

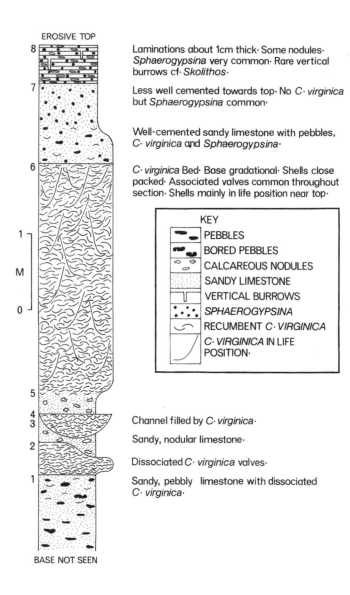

Figure 19.1 Graphic, annotated log of the oyster bank, Round Hill Beds, August Town Formation at Farquhars Beach, parish of Clarendon, south coast of Jamaica (after Littlewood & Donovan, 1988, text-fig. 3). Widths of units indicate how beds have weathered relative to each other at this locality. See also Figure 44.1.

preserved. Assemblages comprising a single species are called monospecific, although it is much more usual for them to be polyspecific.

References

Davis, A.G. & Elliott, G.F. (1957) The palaeogeography of the London Clay seas. *Proceedings of the Geologists' Association*, **68**: 255–277.

Donovan, S.K. (1991) The taphonomy of echinoderms: calcareous multi-element skeletons in the marine environment. *In*: Donovan, S.K. (ed.) *The Processes of Fossilization*: 241–269. Belhaven Press, London.

Goodfriend, G.A. (1993) The fossil record of terrestrial mollusks in Jamaica. *In*: Wright, R.M. & Robinson, E. (eds) Biostratigraphy of Jamaica. *Geological Society of America Memoir*, **182**: 353–361.

Littlewood, D.T.J. & Donovan, S.K. (1988) Variation of Recent and fossil Crassostrea in Jamaica. *Palaeontology*, **31**: 1013–1028.

Locatelli, E.R., Briggs, D.E.G., Stemann, T., Portell, R.W., Means, G.H., James-Williamson, S.A. & Donovan, S.K. (2018) Leaves in marine turbidites illuminate the depositional setting of the Pliocene Bowden shell beds, Jamaica. *Geology*, **46**: 131–134.

Pickerill, R.K., Mitchell, S.F., Donovan, S.K. & Keighley, D.G. (1998) Sedimentology and palaeoenvironment of the Pliocene Bowden Formation, southeast Jamaica. *Contributions to Tertiary and Quaternary Geology*, **35**: 9–27.

Rudwick, M.J.S. (2014) *Earth's Deep History: How it was discovered and why it matters*. University of Chicago Press, Chicago.

CHAPTER 20

PALAEOECOLOGY 3: GETTING MORE INFORMATION FROM THE BED

I trust this chapter, certainly more theoretical than practical, will provide useful background. In chapter 18 I made the observation that much of what can be witnessed or measured directly by the ecologist can only be inferred by the palaeoecologist. That is not to denigrate our inferences, which improve as we gain experience. Some palaeoecological data may be obtained by, for example, precise isotope analysis, as outlined below. I include this because it is interesting and relevant to know that such measurements are possible, but they remain expensive and somewhat outside the scope of the kitchen table laboratory. Certainly, it is not a tool for the amateur or undergraduate, but that is no reason not to note published data related to subjects of interest to you.

But I remember the old saying that an engineer can do for fifty pence what anyone else can do for a pound. Within the limits of the facilities that you have to hand, and with practice and experience, you will be doing fifty pence palaeoecological analyses as a matter of course.

Palaeotemperature

Palaeotemperature analysis is a branch of palaeoecology that attempts to interpret part of the ancient physical environment by indirect methods. The nature of ancient sedimentary rocks and fossils can often be used to obtain a qualitative estimate of ancient temperatures. For example, by analogy with recent environments, it is surely true that ancient, shallow-water reefs would have been tropical in origin.

Quantitative estimates of ancient temperatures may be obtained by the laboratory analysis of stable oxygen isotopes retained by certain fossil shells. Oxygen has three naturally occurring isotopes: ^{16}O, ^{17}O and ^{18}O, which occur in a natural ratio at the present day of 99.76%/0.04%/0.20%. In stable isotope studies the ratio $^{18}O/^{16}O$ is used, because ^{18}O is five times more abundant than ^{17}O and it is easier to measure the contrast in atomic weights between ^{18}O and ^{16}O than between ^{17}O and ^{16}O (Brenchley & Harper, 1998, pp. 43–50).

In stable isotope studies the ancient sample of interest is compared with a standard of known isotopic ratio within a mass spectrometer. One such standard uses carbon dioxide released when the Cretaceous belemnite *Belemnitella americana* (Morton) from the Peedee Formation, the Carolinas, USA, is dissolved in 100% phosphoric acid. The $^{18}O/^{16}O$ ratio of the sample is measured and compared with the standard. The deviation of the sample from the standard is expressed in parts per thousand, that is, per mille, ‰. For ocean water, $^{18}O/^{16}O$ varies between 3‰ below standard at 30°C to 3‰ above standard at 5°C. Differences are thus small. The following difficulties are inherent in this method.

1. $^{18}O/^{16}O$ ratios of sea water are both temperature and salinity dependent. This problem is overcome by using fossils that lived in an open ocean environment, where it is assumed that salinity has remained constant at about 35‰.

2. The $^{18}O/^{16}O$ ratio can only be used if the skeletal calcite was secreted in equilibrium with the surrounding sea water.

3. It is necessary to know the original $^{18}O/^{16}O$ ratio of sea water. This is not only influenced by variations of temperature and salinity, but also by glaciations. Ice is enriched in ^{16}O, so during an 'ice age' sea water is enriched in ^{18}O. In practice, this results in calculations of palaeotemperature being relative, not absolute. Thus, a calculated palaeotemperature of 18°C is more likely to indicate a real temperature of 16°C to 20°C rather than one of less than 5°C.

4. Some organisms only grow at certain temperatures or certain seasons. A pelagic organism, which pursued its optimal environmental conditions in life, will be of greater utility for detecting seasonal variations in temperature than a member of the sessile benthos.

5. Diagenetic changes can alter $^{18}O/^{16}O$ ratios. It is essential to know the preservation history of a fossil before examining its isotopic data. Corals, gastropods and ammonites are aragonitic; echinoderms and some algae secrete a high-magnesium calcite skeleton. All of these commonly undergo chemical modifications during preservation. Planktonic foraminifers, brachiopods and certain molluscs are the best sources of oxygen isotopic data; all have skeletons of stable low-magnesium calcite. Best of all are the belemnites, whose massive skeletons are particularly resistant to diagenetic change.

Palaeodepth

The usual methodologies for estimating marine palaeodepth are dependent upon uniformitarian principles. By observing the depth range of an extant organism or group of organisms, it is attempted to extrapolate this data into the fossil record. Obviously such studies are most convincing when they include multiple lines of information.

The scleractinian corals, which are divided into two ecological assemblages, are a group widely used in palaeodepth analysis. This is related to the presence or absence of symbiotic algae, the zooxanthellae, within the tissues of the coral. These assimilate metabolic waste of the coral, produce oxygen (thus aiding coral respiration) and carbon dioxide (assisting in the secretion of aragonite in the coral skeleton). Hermatypic (reef-building) zooxanthellate corals are the principal framework of modern shallow-water reefs, along with calcareous algae. Such shallow-water reefs have a restricted depth, temperature and salinity range at the present day (Ager, 1963, pp. 39–43).

	Range	Optimum conditions
Depth	Surface to 90 m	Surface to 20 m; best at 15 m or less
	(Restricted by photosynthesis of zooxanthellae)	
Temperature	(17—18°C) 22—36°C	25—29°C
	(Regions of high temperature=regions of strong sunlight)	
Salinity		35‰ ± 1‰

Modern, shallow-water coral reefs include a diversity of coral species associated with calcareous algae and a rich invertebrate fauna. They should not be confused with deep-water coral banks, which comprise a few hermatypic species, no calcareous algae and a sparser invertebrate fauna (Horland & Risk, 2003).

This limitation on the depth of coral reefs, controlled by the requirements of zooxanthellate and calcareous algae, is related to the depth of sea water that is penetrated by sufficient sunlight for photosynthesis, the photic zone. This is about 100 m in open sea water. Some extant scleractinian genera or closely related taxa are known from the Upper Triassic, so recognition of the photic zone on the basis of these corals is reasonable for most of the post-Palaeozoic.

Some groups have undoubtedly changed their depth distributions with time. In the Palaeozoic the stalked crinoids were common in both shallow and deep water. Since the mid-Cretaceous stalked crinoids have migrated out of the shallow-water environment and are now limited to water depths over 100 m (Bottjer & Jablonski, 1989; Fig. 20.1 herein). This migration may have been related to increased predation pressures in shallow waters. Modern shallow-water crinoids, the comatulids, essentially lack a stalk. They are capable of swimming and walking with their muscular arms, and are cryptic, hiding in cavities at least during daylight.

The Upper Pliocene Bowden shell bed: a case study of palaeodepth (Fig. 20.2).

The Bowden shell bed of the Bowden Member, south-east Jamaica, is probably the most diversely fossiliferous deposit of the Antillean region, including over 850 fossil taxa and

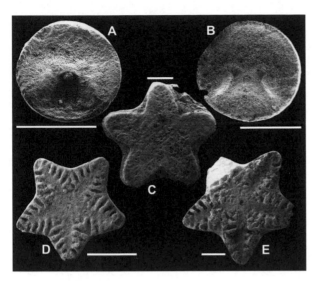

Figure 20.1 Disarticulated ossicles of the Middle Miocene crinoid *Isocrinus* sp. from the Grand Bay Formation, Carriacou, Lesser Antilles (after Donovan & Veltkamp, 2001, fig. 2A–E), a deep water deposit, based on multiple lines of evidence (Donovan *et al.*, 2003). **A, B.** Cirral ossicles, part of the attachment structure. **C.** Distal facet of a nodal columnal, a site of autotomy (self-mutilation). **D, E.** Other articular facets of columnals. All specimens in the Naturhistorisches Museum Basel. Scanning electron micrographs of specimens coated with 60% gold/palladium. Scale bars represent 10 mm.

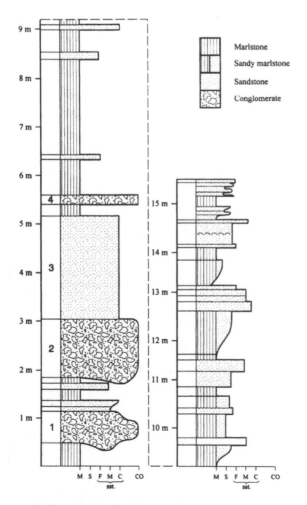

Figure 20.2 Measured section (see Chapter 5) of part of the Upper Pliocene Bowden Member of south-east Jamaica, including the Bowden shell beds (all strata between and including 1–4) (after Pickerill *et al.*, 1998, fig. 2). Key: M = marlstone; S = siltstone; F, M, C are fine-, medium- and coarse-grained sandstone (= sst); and Co = conglomerate.

Marlstone

Sandy marlstone

Sandstone

Conglomerate

Table 20.1 Summary of the principal palaeoenvironmental interpretations of the Pliocene Bowden Member, south-east Jamaica (after Pickerill *et al.*, 1998, table 2; see that publication for details of the papers referred to in the left-hand column).

AUTHOR	PALAEOENVIRONMENT
Vaughan, 1919	< 10 m
Woodring, 1928	< 50 m
Palmer, 1945	~ 30 m
Woodring, 1965	> 200 m
Lagaaij (*in* Rácz, 1971)	12-18 m
Brouwer (*in* Rácz, 1971)	6-21 m
Robinson, 1969a	> 100 m
Goodfriend, 1993	shallow water
Katz and Miller, 1993	upper bathyal (> 200 m)
Robinson, 1994	outer sublittoral (~ 100 m) or deeper

trace fossils. But the unit has also been a conundrum; what depositional depth does it represent (Table 20.1)? This debate goes back at least 100 years and has engendered answers from less than 10 m to greater than 200 m.

Pickerill *et al.* (1998) considered the depth of deposition using multiple lines of evidence, including palaeontology, ichnology (trace fossils) and sedimentology. The fossil assemblage is obviously disharmonious (that is, an allochthonous mix), including land snails and leaves, shallow- and deep-water benthos, fishes, and open ocean plankton. This is indicative of deep-water deposition close to the island shelf from which fossils may be carried into deep water. The sedimentology of the shell beds (1–4 in Fig. 20.2) is in agreement, comprised of coarse-grained rocks (coarse-grained sandstones, conglomerates) representing sediment gravity flows, called turbidity currents, which were generated by tropical storms and earthquakes, and bringing into question whether any of the shelly fossils are *in situ*. This is in contrast to the trace fossils of the marlstones, representing the background sedimentation, and indicative of a hydrodynamically low-energy setting, comparable to known deep-water assemblages.

References

Ager, D.V. (1963) *Principles of Paleoecology*. McGraw-Hill, New York.

Bottjer, D.J. & Jablonski, D. (1989 – for 1988). Palaeoenvironmental patterns in the evolution of post-Paleozoic benthic marine invertebrates. *Palaios*, **3**: 540–560.

Brenchley, P.J. & Harper, D.A.T. (1998) *Palaeoecology: Ecosystems, Environments and Evolution*. Chapman and Hall, London.

Donovan, S.K., Pickerill, R.K., Portell, R.W., Jackson, T.A. & Harper, D.A.T. (2003) The Miocene palaeobathymetry and palaeoenvironments of Carriacou, the Grenadines, Lesser Antilles. *Lethaia*, **36**: 255–272.

Donovan, S.K. & Veltkamp, C.J. (2001) The Antillean Tertiary crinoid fauna. *Journal of Paleontology*, **75**: 721–731.

Horland, M. & Risk, M. (2003) Do Norwegian deep-water coral reefs rely on seeping fluids? *Marine Geology*, **198**: 83–96.

Pickerill, R.K., Mitchell, S.F., Donovan, S.K. & Keighley, D.G. (1998) Sedimentology and palaeoenvironment of the Pliocene Bowden Formation, southeast Jamaica. *Contributions to Tertiary and Quaternary Geology*, **35**: 9–27.

CHAPTER 21

PRESERVATION 1: FOSSILIZATION

Having read so far, you will not be surprised if I state the obvious – that fossils are the mineralized remains of once living organisms that are preserved in the sedimentary rock record. The patterns of preservation that we see in the fossil organisms we collect are the result of the many and varied routes through space and time taken following death. Fossils can be found in almost any type of sedimentary rock, including volcaniclastic deposits. Providing that deformation has not been too intense and the original shell was robust, fossils may also be preserved in metamorphosed sedimentary rocks such as slates.

Fossils fall into two broad groups, which we call body fossils and trace fossils.

A body fossil is a mineralized organism or part of an organism that is preserved in a sedimentary rock following death. Typical examples of body fossils include bones and shells. Body fossils are almost invariably composed of the hard parts of the organism and it is only under rare geochemical conditions that unmineralized ('soft') tissues are preserved. Even if unmineralized tissues are preserved, it is only in rare instances that they remain 'soft': for example, Pleistocene mammoths frozen in permafrost in Siberia. Most preservation of soft tissues involves replacement by minerals, such as phosphates and silicates.

A trace fossil preserves no part of the original organism at all. Rather, a trace fossil is preserved evidence of the activities of an organism. Traces are formed by the living organism, unlike body fossils, but are dependent upon burial and lithification for preservation, like a body fossil. Examples of trace fossils include footprints, trails, borings and burrows (see Chapter 25).

The time between the birth of an organism, its death, its partial or complete decomposition or fossilization, and its eventual discovery as a fossil, is divided into a number of sub-disciplines within palaeontology.

BIRTH		
	Palaeoecology	
DEATH		
	Biostratinomy	)
FINAL BURIAL		) **Taphonomy**
	Diagenesis	)
DISCOVERY		

Taphonomy is the study of the processes of fossilization (preservation), essentially those mechanisms that act on a once-living organism following death. Taphonomy is divided into two sub-disciplines of unequal duration: biostratinomy, which is concerned with those processes that act upon a carcass between death and final burial; and diagenesis, which is the study of changes in the chemistry of the fossil and the surrounding sediment (upon lithification, sedimentary rock). Although diagenesis is principally the domain of the sedimentologist, it is also of great importance to the palaeontologist.

Biostratinomy of individual organisms

It is easier to preserve a shell that is composed of only a single part, such as a gastropod (snail), than it is to fossilize a complex skeleton composed of many separate elements, such as the endoskeleton of a vertebrate or an echinoderm. In order to preserve a complex skeleton as a whole, it is necessary that burial should be early, before the soft tissues that hold the hard parts together begin to rot, or that the carcass is in a low-energy, anoxic environment that discourages scavengers. If the former, burial should be the cause of death, entombing the animal intact.

Once buried, soft parts will tend to rot away. This act of rotting may alter the local chemical environment and could be associated with the formation of diagenetic minerals such as carbonates or chert (see Chapters 7 and 8). Without rapid burial, complex skeletons will become disarticulated, a process that may introduce tens or hundreds of separate skeletal elements into the sedimentological record (Fig. 23.1). In the aquatic environment dissimilar skeletal components will act differently under the ambient hydrodynamic conditions, so that the parts of a skeleton will become separated. The breakdown of skeletal elements involves a number of factors, including disarticulation, abrasion, bioerosion, dissolution, encrustation and fragmentation.

To illustrate the possible biostratinomic pathways that can be followed by a single organism, consider the preservation of three infaunal (burrowing) bivalve molluscs, A, B and C. All are members of the same species, living in similar environments and with the same potential for preservation.

A: death; undisturbed; shell preserved articulated within original burrow.

B: death; exhumation; disarticulation; abrasion + bioerosion + dissolution + encrustation + fragmentation; final burial of separate valves.

C: death; exhumation; disarticulation; abrasion + bioerosion + dissolution + encrustation + fragmentation; shell completely broken down into sedimentary grains.

Path C is essentially that followed by the majority of organisms, with the consequence that the fossil record represents only a sample of past life.

Preservation of communities

Again, some slight repetition of our understanding of palaeoecology is relevant (Chapters 18 to 20). Although fossils of different species are often found preserved in association, true communities of organisms that were associated in life are only preserved together by catastrophic influences, such as being buried alive by a turbidity current or a volcanic ash fall. Such associations of organisms that lived together in life are called life assemblages.

However, most associations of fossil organisms found together in the same bed or on a bedding plane are mixtures from different environments, so that infaunal (burrowers and borers), epifaunal, planktonic and nektonic elements may all be preserved together. Mixing may be due to lateral transport under the influence of water currents, scavengers dragging carcasses, etc.; or vertical transport, due to the action of burrowing organisms reworking the sediment. Such associations of organisms that did not live together in life are called death assemblages.

Diagenesis

(See also Chapter 24). Once buried, and even before burial, the hard skeletons of organisms are subject to diagenetic alteration. Skeletons may be formed of a variety of chemical compounds, but most commonly they are calcareous, phosphatic or siliceous, or a mixture of these (Tucker, 1991, table 4.1). Even within the calcareous shells, a range of diagenetic changes can occur, due to the different chemical and physical state in which calcium carbonate is precipitated. Examples include:

aragonite – scleractinian corals, gastropods, ammonites.

high-magnesium calcite – echinoderms, some algae.

low-magnesium calcite – belemnites, some foraminifers.

High-magnesium calcite (HMC) is less stable than low-magnesium calcite (LMC) and loses its excess magnesium during diagenesis. A HMC skeleton is more soluble than a LMC skeleton. Aragonite is a metastable form of calcium carbonate and is preserved in the rock record only rarely, more commonly being restructured to calcite or dissolved out.

It is not unusual for original shell material to be dissolved away, leaving a cavity (Fig. 21.1A to B), or to be replaced, such as an originally aragonitic shell being replaced by calcite (Fig. 21.1A to C to D). Such replaced shells are called casts. If, however, a shell is dissolved away to leave a cavity in the rock, the details of the surface of the fossil will be preserved in negative relief in the surrounding sedimentary rock as a mould (Fig. 21.1B, C). The impression of the external surface of a fossil in a sedimentary rock is called an external mould; the impression of the internal surface of a fossil in the rock is an internal mould (Figure 21.1; see also Clarkson, 1998, fig. 1.1).

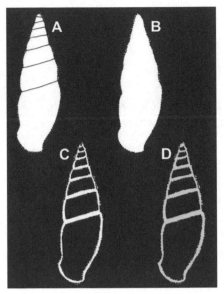

Figure 21.1 Preservation of a gastropod (schematic diagrams based on the Jamaican Pleistocene land snail *Varicella* (*Varicellaria*) *griffithi* (C.B. Adams); Paul & Donovan, 2006, pl. 4, fig. 6), cemented in a brown sandstone. (**A**) The snail shell cemented and unaltered. (**B**) The shell is completely dissolved away by acidic, percolating ground waters. The preservation is an external mould; no internal details are preserved. If infilled by diagenetic minerals, such as calcite or pyrite, the massive resulting specimen, replicating the external features of the shell only, would be a cast. (**C**) The specimen was infilled with sand, which is lithified to sandstone and the shell was then dissolved away. The external mould, similar to (**B**), and the internal mould are separated by a cavity which was formerly the shell. (**D**) Similar to (**C**), but the shell cavity has been infilled by a diagenetic mineral (yellow). This is a cast in which the internal mould is enclosed. Snail about 25 mm in length.

References

Clarkson, E.N.K. (1998) *Invertebrate Palaeontology and Evolution*. Fourth edition. Blackwell Science, Oxford.

Donovan, S.K. (2017) Contrasting patterns of preservation in a Jamaican cave. *Geological Magazine*, **154**: 516–520.

Paul, C.R.C. & Donovan, S.K. (2006) Quaternary land snails (Mollusca: Gastropoda) from the Red Hills Road Cave, Jamaica. *Bulletin of the Mizunami Fossil Museum*, **32** (for 2005): 109–144.

Tucker, M.E. (1991) The diagenesis of fossils. *In*: Donovan, S.K. (ed.) *The Processes of Fossilization*: 84–104. Belhaven Press, London.

CHAPTER 22

PRESERVATION 2: DEATH

Fossils are dead, but how did they die? Obtaining unequivocal evidence of cause of death of a fossil is notoriously difficult. Commonly, it can be done only where some distinctive evidence of, for example, a predator is left on the hard parts of the prey, such as the tooth marks of a shark on a bone or shell (Donovan & Jagt, 2021) or a borehole

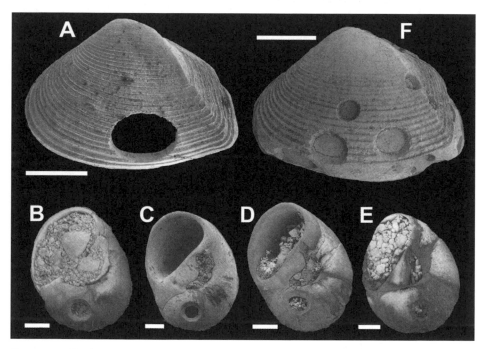

Figure 22.1 Predatory gastropod borings from the Upper Pliocene Bowden shell bed of south-east Jamaica (after Pickerill & Donovan, 1998, pl. 1, fig. 1, pl. 3, figs 4–7, pl. 1, fig. 3, respectively). (**A**) The bivalve *Crassitellites* sp. penetrated by cylindrical *Oichnus simplex* Bromley. This is undoubted predation and almost certainly the trace of a hungry muricid gastropod. (**B–E**) Four shells of the naticid gastropod *Natica castrenoides* Woodring penetrated by bevelled *Oichnus paraboloides* Bromley in identical positions. Not just predation by naticid gastropods, but also probably cannibalism. (**F**) Multiple, incompletely penetrative (= failed predation) borings, *O. simplex*, in the bivalve *Crassitellites* sp. This specimen begs the question, why are there so many 'failed' borings in one valve? All scale bars represent 1 mm.

in a shell made by a predatory gastropod (Fig. 22.1). But it is not common for a fossil to retain evidence of how it dies; many modern predator–prey relationships are known only from direct observation. In the fossil record, it is common either for the cause of death to have left no discernible mark, or a specimen may be too incomplete and the evidence is lost. However, just because evidence for cause of death is rare does not mean that you will not find it.

Predation

Predation is 'box office' palaeontology, as anyone who has seen a movie in the *Jurassic Park* series will know. Many fossil organisms are inferred to be predators on the evidence of morphology and/or on uniformitarian principles. That is, they have structures that suggest a predatory habit, like the jaws of *Tyrannosaurus*, or they are closely related to living predators, such as ancient and modern big cats. Yet few of us will be collecting predatory dinosaurs or Cenozoic big cats. Instead, to illustrate, we will consider an example of predation from among the marine invertebrates.

Marine predatory snails capture their prey by a variety of methods, many of which are specific to a given group and leave no evidence on the hard parts. It is the snails that make predatory boreholes – small round holes of the ichnogenus *Oichnus* Bromley – in their prey organisms, which have a particularly important fossil record. Small round holes are known from throughout the Phanerozoic, but there are two similar forms that are of particular importance in the late Mesozoic and Cenozoic fossil record. Parallel-sided borings are assigned to *Oichnus simplex* Bromley (Fig. 22.1A, F) and are mainly (but not invariably) made by muricid gastropods; more conical or parabolic borings, *Oichnus paraboloides* Bromley, are commonly the spoor of naticid gastropods (Fig. 22.1B–E). Drilling in both groups is a combination of mechanical rasping by the radula ('teeth') and secretion of chemicals that dissolve the shell of the prey. Predation occurs when the gastropod proboscis is inserted through the completed borehole.

Predation may also be non-lethal, that is, the prey survived the attack (Donovan & Jagt, 2021). Autotomy, a form of self-mutilation, is and was an important defensive mechanism in echinoderms, such as crinoids, which are able to lose their appendages to predators and escape. Thus, predation on echinoderms does not have to be lethal. Boreholes into shells may be non-penetrative and thus non-lethal, the predatory snail either having been disturbed or moved on to a more promising prey (Fig. 22.1F).

Disease and parasitism

The range of disease-causing and parasitic organisms in the marine realm is large, from bacteria to fishes. Taking one group of potentially diseased/parasitized invertebrates as our example, the echinoderms, it is only where infestations cause some recognizable deformity in skeletal growth (Fig. 22.2A–D), or in which the infesting organism is

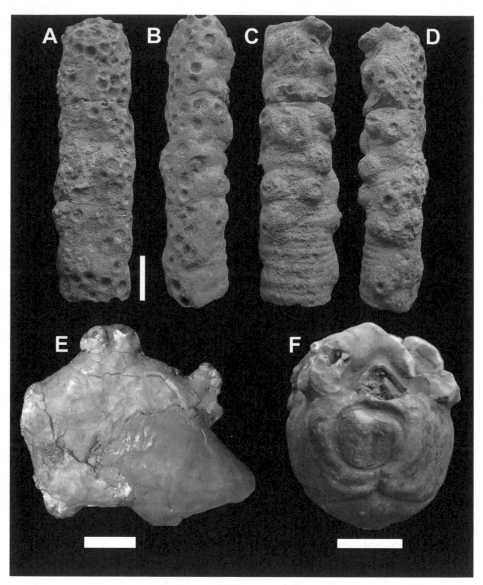

Figure 22.2 **(A–D)** A crinoid pluricolumnal (RGM 791 810) from the Mississippian (Lower Carboniferous) of Salthill Quarry, Clitheroe, Lancashire (after Donovan & Tenny, 2015, fig. 2). The four lateral views are each rotated 90° to the left from the previous image, showing the pattern of pits, *Oichnus paraboloides* Bromley, made by an unknown producer, certainly not a predator. The raised, cyst-like swellings are a reaction by the live crinoid to the pit-former; other pits, on the other side of the specimen, lack swellings and may have been formed after the crinoid died. **(E)** A crinoid theca in oral view, *Platycrinites wachsmuthi* (Wanner) (RGM B16), from the Permian of Timor, with attached platyceratid gastropod conch (after Webster & Donovan, 2012, fig. 1C). **(F)** A crinoid theca in posterior view, *Neoplatycrinus dilatatus* Wanner (RGM B9), showing a circular attachment scar of a platyceratid gastropod immediately beneath the periproct (anus) of the echinoderm (after Donovan & Webster, 2013, fig. 1A). All scale bars represent 10 mm.

preserved, that any indication of disease or parasitism of echinoderms is likely to be recognized in the fossil record.

Palaeozoic crinoids also produced stereom overgrowths in response to encrustation by various non-boring epizoans, such as the tabulate corals; a parasitic relationship is unproven. Stereom overgrowth is a reaction of the crinoid stem to various encrusters, not merely to borers and/or parasites (Fig. 22.2A–D). However, not all crinoid–epizoan associations resulted in calcite overgrowth. For example, the association between platyceratid gastropods and crinoid crowns, in which the mollusc is commonly found surmounting the anal opening of the echinoderm (Fig. 22.2E), produced no overgrowth, but sometimes a 'scar' (Fig. 22.2F), presumably because the snail remained vagile. Similar scars are produced by modern limpets, which return to the same resting position on a rock after every excursion to feed. The crinoid–platyceratid association may have been parasitic, with the snail possibly utilizing the detritus concentrated by the filtration fan of the crinoid, which would thus be denied a portion of its diet. Baumiller (1990) demonstrated that some platyceratids drilled into the crinoid tegmen, which further suggests a parasitic association.

Environmental stress

Marine invertebrates are adversely affected by both desiccation and pollution. Desiccation is important only in shallow-water environments that are subjected to unusually low tides. Pollutants, both man-made and naturally occurring, may cause either mass mortalities or, in smaller concentrations, can be sub-lethal and insidious. For example, abnormalities of echinoid test growth may be produced by some pollutants (Dafni, 1983).

Storms

Storms and the sedimentary processes that they generate may be fatal to shallow-water benthic organisms. Rapid entombment by sediment may bury both epifaunal and infaunal species beyond escape; sediment erosion and washout of (particularly) shallow infaunal burrowers may leave them washed onshore or offshore; and transported pebbles, cobbles and larger clasts may crush thin-shelled organisms. For example, echinoids, asteroids and ophiuroids can be buried alive by thin, storm-generated sediment accumulations. It is the processes of rapid storm-driven sedimentation that can lead to the formation of deposits known as starfish beds, which preserve exquisitely complete specimens of delicate echinoderms and other taxa.

Old age

Taking the echinoids as a typical marine group, they may live between one and 15 years; other echinoderms have durations of less than ten years; but some deep-water species of ophiuroid live up to 20 years. Echinoids that only live for one year may suffer a mass-mortality following spawning. The fossil record of such events, although rarely preserved, will probably appear as a bed dominated by a single species, with all specimens about the same size. In a species that shows sexual dimorphism – that is, the sexes are different in appearance, as in many ammonite species – such a bed may include both smaller and larger specimens, that is, males and females respectively.

References

Baumiller, T.K. (1990) Non-predatory drilling on Mississippian crinoids by platyceratid gastropods. *Palaeontology*, **33**: 743–748.

Dafni, J. (1983) Aboral depressions in the tests of the sea urchin *Tripneustes* cf. *gratilla* (L.) in the Gulf of Eilat, Red Sea. *Journal of Experimental Marine Biology and Ecology*, **67**: 1–15.

Donovan, S.K. & Jagt, J.W.M. (2021). Ichnology of Late Cretaceous echinoids from the Maastrichtian type area (The Netherlands, Belgium) – 4. Shark *versus* echinoid: failed predation on the holasteroid *Hemipneustes*. *Bulletin of the Mizunami Fossil Museum*, **47**: 49-57

Donovan, S.K. & Tenny, A. (2015) A peculiar bored crinoid from Salthill Quarry, Clitheroe, Lancashire (Mississippian; Tournaisian), UK. *Proceedings of the Yorkshire Geological Society*, **60**: 289–292.

Donovan, S.K. & Webster, G.D. (2013) Platyceratid gastropod infestations of *Neoplatycrinus* Wanner (Crinoidea) from the Permian of West Timor: speculations on thecal modifications. *Proceedings of the Geologists' Association*, **124**: 988–993.

Pickerill, R.K. & Donovan, S.K. (1998) Ichnology of the Pliocene Bowden shell bed, southeast Jamaica. *Contributions to Tertiary and Quaternary Geology*, **35**: 161–175.

Webster, G.D. & Donovan, S.K. (2012) Before the extinction – Permian platyceratid gastropods attached to platycrinitid crinoids and an abnormal four-rayed *Platycrinites s.s. wachsmuthi* (Wanner) from West Timor. *Palaeoworld*, **21**: 153–159.

CHAPTER 23

PRESERVATION 3: DISARTICULATION, TRANSPORT AND RESIDENCE

Disarticulation

Many organisms have one-piece hard parts with a high preservation potential, such as solitary and massive colonial corals, gastropods and belemnites. There are also organisms with a two-piece skeleton that, even when disarticulated, are easily identified and reunited with their other half, such as bivalve molluscs and brachiopods. Yet there are many organisms with a complicated, multi-component skeleton that commonly disarticulates after death, such as plants, arthropods, echinoderms and vertebrates. Palaeontologists are used to finding skeletons in pieces and rejoice when a complete trilobite, echinoid or fish is discovered. But dealing with and understanding the fragments of your favourite organisms is a specialization worth pursuing.

You were introduced to the Red Hills Road Cave in Jamaica in Chapter 13. This is extraordinarily rich in disarticulated bones, but lacks any articulated vertebrate skeletons

Figure 23.1 A prehistoric Jamaican stew? Bones, shells and rare arthropod remains of the Late Pleistocene Red Hills Road Cave of Jamaica (image provided by Ms Els Baalbergen). The sample has been processed with a coarse mesh sieve to remove excess sediment and concentrate fossil remains. The large bone to the left is a femur of the extinct Jamaican ibis, *Xenicibis xympithecus* Olson & Steadman.

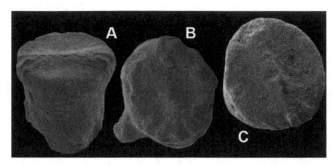

Figure 23.2 Columnals of *Platycrinites* sp. *sensu lato* from the Mississippian (Lower Carboniferous) of Salthill Quarry, Clitheroe, Lancashire, UK (after Donovan, 1997, pl. 11, figs 4–6). Specimens registered in the Natural History Museum, London (prefix BMNH). **(A)** BMNH EE 5233[2], oblique view of high columnal, about 1.55 mm diameter. **(B)** BMNH EE 5231[1], articular facet and tubercle, about 1.71 mm maximum dimension. **(C)** BMNH EE 5236[1], articular facet, about 1.83 mm maximum diameter. Scanning electron micrographs of specimens coated with 60% gold-palladium.

(Donovan, 2017). It is the antithesis of the perfect vertebrate site (Fig. 23.1), yet some of the fragments have now been identified and, rarely, a mammal jaw with teeth is identifiable. So, even working with just bones and teeth can provide solid data. The fossil record of sharks is almost exclusively teeth, but they have been studied in minute detail and the geological history of the group is well known.

I am an echinoderm worker and the individual plates (= ossicles) of the echinoderm skeleton can be recognizable, at least in a broad sense. Structures such as echinoid spines and crinoid columnals are broadly identifiable and, less commonly, can be tied down to family, genus or species. For example, many crinoids have columnals that are merely variations on the theme of a circular outline with a central, circular axial canal. Yet, for example, there are distinctive elliptical columnals common in the Upper Palaeozoic and that are derived from the Family Platycrinitidae Austin & Austin (Fig. 23.2).

Transport

Identifying whether a fossil has been transported or not is rarely simple. We recognize three 'states' of transport history: autochthonous (in life situation); parautochthonous (any transport is strictly local); and allochthonous (transported). Fossils of free-swimming and floating organisms, such as most ammonoids and fishes, are allochthonous fossils, rather obviously: after death, they will sink to the sea floor, perhaps after a period of floating and transport by currents. But what of benthic invertebrates? A Palaeozoic limestone preserving abundant crinoid columnals and pluricolumnals (portions of stem) may have the appearance of transport, but a dead crinoid may disarticulate in days. What appears to be an allochthonous assemblage of crinoid plates may merely be one or more collapsed individuals.

Disarticulated infaunal bivalves may be informative indicators of transport. Infaunal bivalves preserved in life position are easily recognized. If dead shells are reworked and disarticulated, count the number of left and right valves of a common species (you can

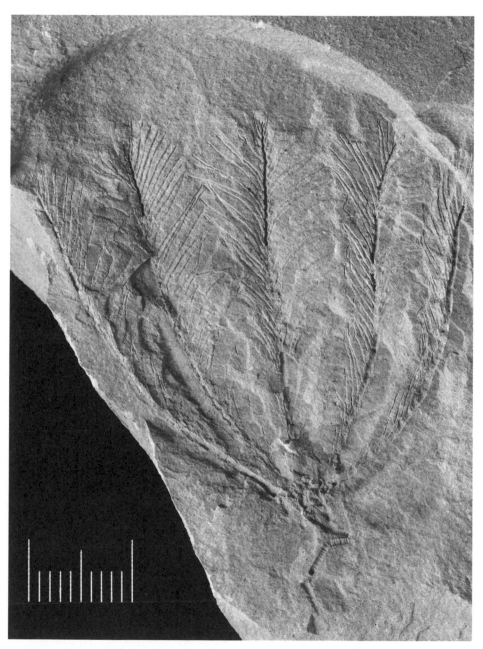

Figure 23.3 *Euptychocrinus longipinnulus* Fearnhead *et al.*, NMW 2019.27G.1, holotype, natural external mould (after Fearnhead *et al.*, 2020, fig. 2). Image by Mr. Harry Taylor (Photographic Unit, the Natural History Museum, London). Whitened with ammonium chloride (see Chapter 28). Scale in mm. See also Figure 32.1.

practise such an analysis on a local beach). If the number of lefts and rights are about equal, then they may not have been transported far following disinterment, if at all. But if either left or right valves are numerically dominant, the chances are that there has been both transport and sorting. This is because the left and right valves of all infaunal bivalves are mirror images of each other and their hydrodynamic behaviours are different; that is, right and left valves move in different directions in waves and water currents (see also Chapter 11).

Articulated crowns of early Palaeozoic crinoids are rare fossils in the British Isles. These would intuitively be assumed to be allochthonous, but the opposite may be true. Consider the Lower Silurian (Llandovery) *Euptychocrinus longipinnulus* Fearnhead *et al.*, 2020 (Fig. 23.3). The contrasting preservation of stem vs. arms in this specimen is unexpected. In extant crinoids, it is the pinnules – that is, the fine, feather-like branches from the arms – that start to disarticulate first upon death. Yet *E. longipinnulus* has a complete preserved pinnulation with cover plates; burial must have been the cause of death, by comparison with numerous other specimens from throughout the Phanerozoic. Yet the stem only preserves part of the proxistele (= proximal stem) which is broken into four discrete sections, the most proximal still attached to the base of the cup. No other body fossils were preserved at this location. Most likely, this crinoid was disturbed by, for example, a storm and tried to escape down-current by autotomizing (breaking off by self-mutilation) the more distal stem. The breakage of the proxistele into discrete lengths, each containing about the same number of columnals, is likely explainable by the 'broken stick' model of Baumiller & Ausich (1992; Donovan, 2021), lengths of fragments being controlled by the pattern of ligamentation. The stem was mostly autotomized and the crinoid was carried away by the storm current, but, instead of being carried to a more benign environment, it was buried alive, hence the exquisite preservation of the arms and pinnules.

Residence

An organism dies and then it is buried. The enquiring palaeontologist should ask if the duration of time represented by 'and' in the previous sentence can be determined? The case of *Euptychocrinus* in the previous section of this chapter is an extreme example; as the crinoid was buried alive, its residence time on the surface was zero in whatever units of time you choose. Or, if you would like a more prosaic answer, it all happened in a flash one Thursday afternoon.

My favourite seafloor 'residents' are big, chunky shells such as the large shells of tropical strombid gastropods, oysters and holasteroid echinoids, which are locally common in Upper Cretaceous chalk and other limestones. Oysters may be articulated or disarticulated. Even in life, oysters may support various encrusting organisms, not least juvenile oysters (Donovan & de Gier, 2020). Barnacles on the inside of a valve must have encrusted the surface after the death of the oyster, but beware: some encrusting organisms, most notably acorn barnacles (*Balanus* sp.), invade the inside of a shell as soon as the soft tissues rot away and before the ligament linking the valves has broken (Donovan *et al.*, 2014); that is, very soon after death.

References

Baumiller, T.K. & Ausich, W.I. (1992) The Broken-Stick model as a null hypothesis for crinoid stalk taphonomy and as a guide to the distribution of connective tissue in fossils. *Paleobiology*, **18**: 288–298.

Donovan, S.K. (1997) Comparative morphology of the stems of the extant bathycrinid *Democrinus* Perrier and the Upper Palaeozoic platycrinitids (Echinodermata, Crinoidea). *Bulletin of the Mizunami Fossil Museum*, **23** (for 1996): 1–27.

Donovan, S.K. (2017) Contrasting patterns of preservation in a Jamaican cave. *Geological Magazine*, **154**: 516–520.

Donovan, S.K. (2021) Train crash crinoids revisited. *Lethaia*, **54**: 1-3.

Donovan, S.K., Cotton, L., Ende, Conrad van den, Scognamiglio, G. & Zittersteijn, M. (2014) Taphonomic significance of a dense infestation of *Ensis americanus* (Binney) by *Balanus crenatus* Brugière, North Sea. *Palaios*, **28** (for 2013): 837–838.

Donovan, S.K. & de Gier, W. (2020) The oyster shell as habitat: encrusters and borings. *Deposits*. depositsmag.com/2020/05/25/the-oyster-shell-as-habitat-encrusters-and-borings.

Fearnhead, F.E., Donovan, S.K., Botting, J.P. & Muir, L.A. (2020). A Lower Silurian (Llandovery) diplobathrid crinoid (Camerata) from mid-Wales. *Geological Magazine*, **157**: 1176–1180.

CHAPTER 24

PRESERVATION 4: BURIAL AND DIAGENESIS

Burial

In its slow (or not so slow) walk from life to death to final burial, we have arrived at the terminal station. Burial is not a once-and-for-all event. Many skeletal remains may be buried and exhumed before final burial. Others, like *Euptychocrinus* in Chapter 23, were buried just once, but catastrophically, that being the cause of death. Being buried alive means different things to different groups of organisms. Our Silurian crinoid had no ability to escape from even shallow burial, even a few millimetres. Contrast this with another, but very different, echinoderm group, the post-Palaeozoic heart urchins (spatangoids) (Fig. 24.1), members of which were, and are, superbly adapted to a subterranean existence (Smith, 1984).

Rather obviously, organisms that lived infaunally were likely to already be buried at the time

Figure 24.1 Heart urchin *Schizaster* sp. cf. *S. americanus* (Clark & Twitchell), University of Alberta P1670, Middle Miocene, Cayman Formation, Grand Cayman (after Donovan *et al.*, 2016, fig. 3B). Internal mould of irregular echinoid, apical view. Scale bar represents 10 mm. Specimen whitened with ammonium chloride.

of death. This group of organisms is diverse, including heart urchins and other burrowing echinoids, but also burrowing and boring bivalves, certain gastropods such as naticids (Fig. 22.1B-E) and the blind trinucleid trilobites. These were all vagile organisms, actively digging or boring.

Contrast these with one Palaeozoic group that was passively infaunal, the productid brachiopods (Fig. 24.2). Like other articulated brachiopods, the concavo-convex productids lacked any mechanism for active burrowing. Rather, by adaptations such as massive shells and stabilizing spines they were committed to remaining sessile. They would have been easily buried, but retained a contact with the external environment by extending the commissure above the surface like a snorkel. Thus, they retained a

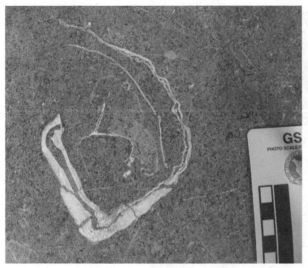

Figure 24.2 An antero-posterior section of a particularly large productid brachiopod in Mississippian limestone, a facing stone in the Nachtegaalstraat, Utrecht, the Netherlands (after Donovan & Harper, 2018, fig. 3). The image has been rotated through 180° to show the brachiopod in life position. Scale in cm.

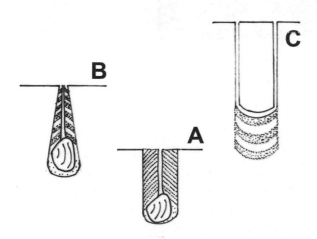

Figure 24.3 Vertical burrows showing differing responses to sedimentation (redrawn after Häntzschel, 1975, figs 14.1c, 14.1a, 16.3, respectively). **A.** Rapidly burrowing deeper by bivalve in response to erosion; spreite above shell equal to shell width. **B.** Slowly burrowing deeper by bivalve in response to growth; width of spreite above shell equal to width of shell at time of formation. **C.** *Diplocraterion* Torell burrow (U-shaped) digging up in response to surface sedimentation, leaving spreite underneath.

contact with the bottom waters, drawing in currents rich in microplankton and dissolved oxygen.

The response of burrowing organisms to sedimentation and erosion may be preserved by their trace fossils (Häntzschel, 1975, pp. W27–W32). Erosion of the sea floor will bring an infaunal organism to less than its optimum sediment depth. The response will likely be to dig deeper, leaving distinctive sedimentary structures (Fig. 24.3A). Similar, but distinct, structures are produced in examples where the shell lived in an area of stable sediment and dug deeper as it grew bigger (Fig. 24.3B).

Conversely, in an area where net sedimentation occurs, an infaunal burrower will need to move up in the sediment column to prevent being buried too deep to live (Fig. 24.3C). Such burrows are called escape structures.

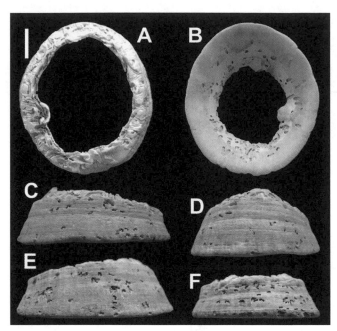

Figure 24.4 *Patella vulgata* Linnaeus (specimen in Naturalis Biodiversity Center, Leiden), Recent, incomplete and densely infested by U-shaped borings *Caulostrepsis taeniola* Clarke (after Donovan, in press, fig. 2). Seen in apical, basal (apertural) view, left lateral, anterior view, right lateral and posterior views (A–F, respectively). Specimen uncoated. Scale bar represents 10 mm.

While on holiday, I collected a Recent limpet preserved in an unusual manner (Fig. 24.4). It has largely lost its apex, perhaps exclusively due to the weakening action of an infestation of boring annelid worms. It was on Queen Victoria's private bathing beach, Isle of Wight, which is open to all visitors to Osborne House, East Cowes (see Chapter 49). The specimen, probably *Patella vulgata* Linnaeus, has lost perhaps two-thirds of the apex. The exposed thickness of the shell has been densely infested by U-shaped borings, *Caulostrepsis taeniola* Clarke. The likely scenario for its post-mortem deterioration is that the dead shell would have been stable on the sea floor in an aperture-down orientation, buried by sediment covering about the lower one-third of the shell.

The apex would have presented a benthic island and was infested by *C. taeniola* Clarke. These borings are common in the remnant of the shell; other infestors may have been present on the lost apex, but have left no evidence. The residence time in this position was long enough for the entire apex to be lost due to bioerosion. That is, even partial burial may leave a unique preservational signature.

These varied examples demonstrate that careful observations of fossil specimens may enable you to determine a history of burial. Consider, as one more example, a boring bivalve. These may be preserved within their boring, body and trace fossil in close association. (But beware – some non-boring organisms will invade a discarded boring; see, for example, Newall, 1970). If boring bivalves are found released from their borings, then they have been eroded out; perhaps they bored in a friable mudrock or peat (Donovan, 2013). If the external surface of the valve(s) is encrusted, then this must have occurred after it was released from the burrow (Donovan, 2020, fig. 5).

Diagenesis

Diagenesis encompasses all the processes affecting a sediment and its contained fossils from deposition until metamorphism ... The diagenesis of fossils depends primarily on the original mineralogy and on the diagenetic environment (Tucker, 1991, p. 84).

The time available for diagenesis is potentially vast – far longer than the residence time of any shell or bone on the sea floor. So, what can happen, will happen. The skeletons of animals are commonly composed of calcium carbonate (calcite and aragonite), silica or phosphate. All are physically 'hard', yet under the right conditions of pH (acidity and alkalinity) all may be dissolved away. For example, aragonite, which forms the skeletons of scleractinian corals, gastropods and some bivalves, and a range of other organisms, is less stable than calcite and is commonly dissolved away and/or replaced. Even some forms of calcite undergo chemical modification. Echinoderms (and certain other groups: Tucker, 1991, table 4.1) have a skeleton of high (but not very high) magnesium calcite. Echinoderm skeletons lose their magnesium during their first million years or so of burial and re-equilibrate as low magnesium calcite. Calcitic brachiopods, certain foraminifers and belemnites have a low-magnesium calcite skeleton that is chemically stable and makes them a favourite for stable isotope geochemistry. Replacement minerals can be exotic, depending on the diagenetic environment, as indicated by Tucker, above, and can include pyrite and precious opal, for example.

References

Donovan, S.K. (2013) A distinctive bioglyph and its producer: Recent *Gastrochaenolites* Leymerie in a peat pebble, North Sea coast of the Netherlands. *Ichnos*, **20**: 109–111

Donovan, S.K. (2020) Fossils explained 78: Never bored by borings: an ichnologist in Margate. *Geology Today*, **36**: 232-235.

Donovan, S.K. (in press). Taphonomy of a limpet. *Ichnos*.

Donovan, S.K. & Harper, D.A.T. (2018) Urban geology: productid brachiopods in Amsterdam and Utrecht. *Deposits*, **56**: 10–12.

Donovan, S.K., Jones, B. & Harper, D.A.T. (2016) Neogene echinoids from the Cayman Islands, West Indies: regional implications. *Geological Journal*, **51**: 864–879.

Häntzschel, W. (1975) Trace fossils and problematica. Second edition (revised and enlarged). *In*: Teichert, C. (ed.) *Treatise on Invertebrate Paleontology, Part W, Miscellanea, Supplement 1.* Geological Society of America and University of Kansas, Boulder and Lawrence.

Newall, G. (1970) A symbiotic relationship between *Lingula* and the coral *Heliolites* in the Silurian. *In*: Crimes, T.P. & Harper, J.C. (eds) *Trace Fossils. Geological Journal Special Issue*, **3**: 335–344.

Smith, A.B. (1984) *Echinoid Palaeobiology*. George Allen and Unwin, London.

Tucker, M.E. (1991) The diagenesis of fossils. *In*: Donovan, S.K. (ed.) *The Processes of Fossilization*, pp. 84–104. Belhaven Press, London.

CHAPTER 25

TRACE FOSSILS

Trace fossils are biogenic or bioerosional sedimentary structures, and as such may be regarded as a common interest of sedimentologists and palaeontologists alike. A trace fossil does not preserve part of the skeleton (= body fossil) of the producing organism. Rather, a trace fossil is the result of the activities of an organism interacting with the substrate. While it is commonly possible to determine the affinities of body fossils by comparison with extant organisms, it may be impossible to determine the precise (or even general) identity of an organism that generated a particular trace fossil, with some notable exceptions, because:

a) Any given species is capable of generating a diversity of trace fossils.
b) Conversely, but in many examples, a given morphology of trace fossil may have been produced by any one of a variety of taxa.

You can demonstrate point (a) easily next time you spend a day on the beach or in the snow. See how many different traces you can make in the sand or snow, just using your own feet. You can go for a walk, your separate tracks combining to form a trackway. But if you drag your feet they will form two parallel trails. What if you hop on one leg? That is just the start of what you can do using your two feet; you also have two hands and the rest of your body to get in on the act.

Point (b) is confirmed by vertical U-shaped burrows that are commonly found in shallow marine environments under high-energy conditions at the present day, and can be produced by a variety of annelid worms and arthropods. Similar burrow morphologies have a broad range in space, but the extant organisms that produce them are limited geographically. Such U-shaped burrows are known from similar palaeoenvironments back into the Mesozoic and Palaeozoic (Fig. 25.1A), but there is little evidence of the producing organisms in ancient examples. As in modern examples, the producers presumably lacked mineralized hard parts and thus had a low potential for preservation. A variety of organisms belonging to different major groups probably generated U-shaped burrows in the ancient past. We can speculate on the basis of their trace fossils, but no more than that, unless the burrower is preserved *in situ*.

The common forms of trace fossil are introduced below.

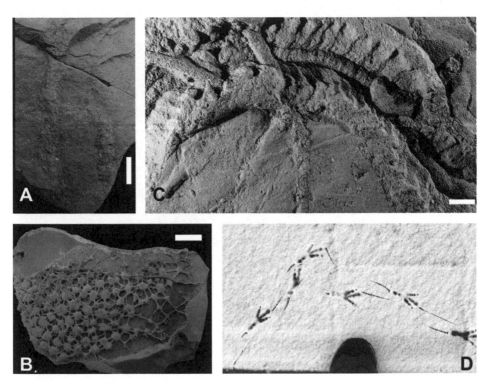

Figure 25.1 Various traces, ancient and modern. **A.** The U-shaped *Arenicolites* isp. presented in its inferred life position (after Donovan *et al.*, 2020, fig. 3B). Natural History Museum, London (NHM), number TF 100019. Ketton Quarry SSSI, East Midlands, England; Whitby Mudstone Formation; Jurassic, upper Lias, Toarcian. Collected from float. **B.** Flint preservation of a complicated boring. Upper Cretaceous *Entobia cretacea* (Portlock), NHM S.9015, a clionaid sponge boring preserved in flint and collected as a reworked specimen in chalk drift, Croydon, Surrey, England (after Donovan & Fearnhead, 2015, fig. 1). The sponge bored in a valve of a dead *Inoceramus* sp. bivalve. The boring was subsequently infilled by flint during diagenesis, but the shelly substrate was then dissolved away. **C.** Trail *Scolicia prisca* de Quatrefages, Naturalis Biodiversity Center, Leiden, number RGM 188 697, from the Paleogene Richmond Formation of Jamaica (after Donovan *et al.*, 2005, fig. 1A). This is a general view of the more sinuous, and better preserved, part of specimen; note that it is cross-cut by the burrows *Planolites* isp. cf. *P. beverleyensis* (Billings) in the left side of this view. **D.** Recent bird trackway in the snow, probably made by a thrush (after Donovan & Fearnhead, 2014, fig. 1B). Scale bars in A–C represent 10 mm.

Burrows are produced by organisms digging down into and through a soft substrate. Burrows can be simple, either straight or U-shaped, and unbranched, or complex and branched (Fig. 25.1A). They may be horizontal, vertical or any angle in between, or a network in three dimensions.

Borings are morphologically similar to burrows, but result from the producing organism cutting into a hard substrate, such as rock, bone or wood (Fig. 25.1B). It is therefore essential to identify the original state of the substrate before it is determined whether it has been bored or burrowed. Note that both burrows and borings occur within substrates. In contrast, trails and tracks are surface traces.

Trails are continuous grooves produced during locomotion by an animal having part of its body in contact with the surface of the substrate (Fig. 25.1C). A typical trail producer might be a gastropod or worm. Trails can vary from simple to highly complex, the latter generally produced by organisms that were grazing on the sediment surface. The more closely packed the folds of such trails, the more efficient the feeding behaviour. Grazers may infill their trails with a discontinuous or continuous ribbon of sediment produced by defecation.

Tracks are the impressions left in the underlying sediment by the individual feet of an organism walking with its body raised above the substrate, such as your footprint in the sand or snow. A succession of tracks is called a **trackway** (Fig. 25.1D). Trackways are mainly produced by arthropods and terrestrial tetrapods.

Coprolites are preserved faecal material and may be diagnostic or preserved in association with the producing organisms. For example, sharks produce distinctive spiral coprolites. The main uses of coprolites are in determining the diet of ancient organisms and, in some late Pleistocene terrestrial tetrapods, such as Shasta ground sloths, for radiocarbon dating.

Systematics of trace fossils

Trace fossils are classified in a manner that is superficially similar to Linnean systematics. Ichnospecies are included in ichnogenera, each of which is Latinized (for example, *Oichnus simplex* Bromley). The naming of trace fossils in this form is an accident of history, from the nineteenth century when burrows were at first misidentified as fossil plants (Osgood, 1975).

However, these names do not form part of a nested hierarchy; although some authors use an ichnofamily concept to group together particular trace fossils, they have no biological validity. Further, the names of trace fossils are not meant to imply relationships to particular groups of organisms. Do not think of them as organisms— they are sedimentary structures with fancy names. These names are just convenient labels that group together morphologically similar traces without implying any genetic relationship between the producing organisms (although such a relationship may, nevertheless, exist).

Behavioural classification of trace fossils

Trace fossils are interpreted as having been produced commonly, but not invariably, during one or more of the following activities:

Dwelling – **Domichnia**

Feeding – **Fodinichnia**

Grazing – **Pascichnia**

Crawling – **Repichnia**

Resting – **Cubichnia**

Predation – **Predichnia**

This list is not exhaustive, but includes the most common behavioural categories represented by invertebrate trace fossils. This classification is based on the observation that different groups of animals will produce similar traces under similar environmental conditions.

Utility of trace fossils in palaeontology

In palaeontology, trace fossils can be used to give a broad indication of the morphology of otherwise unknown organisms, particularly those that lacked mineralized hard parts (Fig. 25.1A), and the rarely preserved organs of known organisms, such as the feet and legs of trilobites. The stratigraphic ranges of trace fossils may also supply important information about the evolution of certain groups. For example, the furrowing traces called *Cruziana* are commonly interpreted as produced by the locomotion of trilobites. However, they are known from Cambrian deposits that pre-date the first trilobites. This may suggest an early evolutionary phase when the trilobite exoskeleton was not mineralized: indeed, unmineralized trilobites are known from the Middle Cambrian Burgess Shale of British Columbia (Gould, 1989, pp. 164–167). However, it also indicates that *Cruziana* may be produced by other, bilaterally symmetrical organisms that are not trilobites; significantly, *Cruziana* is known from the post-Palaeozoic, after the demise of the trilobites (Donovan, 2010). Similarly, interpretations of the functional significance of trace fossils may provide important information about the gross morphology of ancient, unknown taxa.

References

Donovan, S.K. (2010) *Cruziana* and *Rusophycus*: trace fossils produced by trilobites ... in some cases? *Lethaia*, **43**: 283–284.

Donovan, S.K. & Fearnhead, F.E. (2014) The nature of trace fossils. *Deposits*, **39**: 38–43.

Donovan, S.K. & Fearnhead, F.E. (2015) Exceptional fidelity of preservation in a reworked fossil, Chalk drift, South London, England. *Geological Journal*, **50**: 104–106.

Donovan, S.K., Renema, W. & Pickerill, R.K. (2005) The ichnofossil *Scolicia prisca* de Quatrefages from the Paleogene of eastern Jamaica and fossil echinoids of the Richmond Formation. *Caribbean Journal of Science*, **41**: 876–881.

Donovan, S.K., Strother, P. del & Ewin, T.A.M. (2020). A rare and unusual trace fossil from the Lower Jurassic (Lias Group) of Ketton, East Midlands. *Proceedings of the Yorkshire Geological Society*, 63: 43–46.

Gould, S.J. (1989) *Wonderful Life: The Burgess Shale and the Nature of History*. W.W. Norton, New York.

Osgood, R.G., Jr. (1975) The history of invertebrate ichnology. *In*: R.W. Frey (ed.) *The Study of Trace Fossils* (pp. 3–12). Springer-Verlag, New York.

WORKING ON YOUR COLLECTION AT HOME

CHAPTER 26

STORAGE

In October 2018 I gave a lecture to the GeoLancashire Group of the Geologists' Association in Clitheroe. The subject of the talk was the geology of Jamaica, but the audience came armed with crinoids from their own collections for me to identify. So, a one hour talk was followed by an hour of crinoid identification and discussion; I had little voice left at the end!

Brian Jeffery came with a big slab of limestone. It had a patchy covering of moss, and woodlice escaped when he put it down on the table. Apparently, Brian's wife wouldn't let rocks accumulate in the house, so the specimen had been in his garden. I asked Brian to clean the slab in a bucket of dilute bleach, and next time that I saw it I was suitably impressed. It was a fine specimen of *Woodocrinus macrodactylus* de Koninck, known only from one quarry, now long infilled, in North Yorkshire. It was an exciting specimen to identify (Donovan & Jeffery, 2020; Fig. 26.1 herein).

Figure 26.1 Rescued from the garden. *Woodocrinus macrodactylus* de Koninck and a crinoid sp. indet. (towards upper left), formerly in the private collection of Brian Jeffery, but now presented by him to the Natural History Museum, London, registration number EE16661 (after Donovan & Jeffery, 2020, fig. 1). The best-preserved specimen of *W. macrodactylus* is right of centre; disarticulated arms are above this and to its left. Specimen uncoated, photographed in natural light. Scale bar represents 10 mm.

So, my obvious first comment is not to keep your prize specimens in the garden! Most specimens will happily survive the temperatures and humidity of your house, so they can live indoors with you. If not, many people use somewhere protected like a garden shed or a corner of the garage.

Each specimen or related group of specimens deserves its own tray or box. (An example of a related group of specimens might be a collection of starfish ossicles from one bed, and probably representing one species or even one individual.) A lid will protect specimens from dust, but it is desirable that you can see the contents without having to remove any covering. So, if you use boxes, the lid should preferably have a window of transparent glass or plastic. For example, I work in a museum, Naturalis in Leiden, which prefers clear plastic boxes with clear plastic lids (Fig. 26.2A). Plastic bags are a possible alternative, but a specimen in a box is more accessible. If a specimen is fragile it will need special consideration (see also Chapter 33), such as extra padding to stop it rolling about. This is more than a trivial consideration; if your collection is in a drawer, then every time it is pulled out or pushed in, every specimen in every box will roll about and, potentially, be damaged.

Numbering and labelling are important (Fig. 26.2B, C). If your collection is small, you can probably carry the information concerning every one of your fossils in your head, but what if it grows to hundreds or thousands of specimens? Each specimen in your collection should have a unique number. This can be written on a small piece of paper,

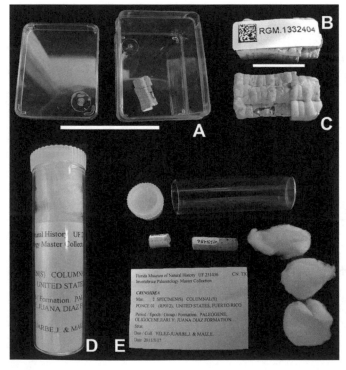

Figure 26.2 Storing specimens. **A.** Clear plastic box (55 x 43 x 30 mm, with lid) and specimen (as **B, C**). Scale bar represents 50 mm. **B, C.** Two views of a specimen – the Paleogene crinoid *Isselicrinus* sp. from Tierra Del Fuego – with its number attached. Naturalis Biodiversity Center, Leiden, number RGM.1332404. Scale bar represents 10 mm. **D, E.** A plastic tube, specimen and label, both as stored (**D**) and with all components separated out (**E**). A Lower Oligocene crinoid, *Isselicrinus* sp., from Puerto Rico; Florida Museum of Natural History, Gainesville, number UF 231456.

preferably of archival quality and written in permanent ink, and glued on. The glue should be soluble so that the label can be removed if necessary, for example, for photography. Alternately, a number can be written on a large specimen using a fine-tipped permanent marker. A register of these numbers is a logical adjunct, recording relevant data. Your register can either be a computer document or a hard copy in a suitable notebook. The latter has the advantage of ease of access and can be stored with your collection, but a computer document is more easily searched. Labels in the box and accompanying the specimen will duplicate some, probably all of the information from the register, but the latter will have the potential to record more, to be more comprehensive.

Recording and duplicating your information about each and every specimen is more than a trivial concern.

Doubtless, Benjamin Walworth Arnold (1865–1932)… knew where he had bought and collected the fossil echinoids [from Jamaica] described in Arnold and Clark (1934, p. 139), but he died before he had informed his co-author. Similarly, there are a number of drawers of specimens in the Sedgwick Museum [University of Cambridge] stores … labelled 'do not unpack without speaking to Barrie Rickards'. Unfortunately, Professor Rickards passed away several years ago … (Donovan & Riley, 2013, p. 509).

But what of the labels? What should be recorded? Essentially, a label should record the 'vital statistics' of a specimen. A specimen is numbered, all its details appear in your register, but it is convenient to duplicate important data and store it with the specimen. Important data includes specimen number; specimen identification; locality (if there is space, with grid reference and/or GPS readings, if available); and stratigraphy. The name of the collector (if it wasn't you) ought to be added. Details of when collected may be relevant and will enable cross-referencing with your field notebook. From this list, you will have deduced that writing small and with a fine pen are skills to be developed. Alternately, labels in small print can be manufactured by your word-processing program.

Storing your neatly-organized specimens in boxes is a different problem. Ideally, you will want your collection arranged in some sort of drawer(s) so that, when pulled out, you can see your specimens arrayed before you. As your collection grows, you will need more and yet more drawers, so some sort of cabinet will be ideal. If you lack such a cabinet, then something suitable might be found by haunting shops that deal in second-hand furniture. Suitable cabinets are advertized in the palaeontological literature, but these are aimed more at museums and universities, and may be expensive. Before your collection grows too large, ensure that your floor is strong enough to support a cabinet full of rocks.

So far I have considered what might be called 'average-sized' specimens, as big as your fist or smaller, but what of the big and tiny? 'Big' might include an *Iguanodon* footprint or a mammoth tooth. You will need a big, robust box – probably made of wood for the footprint – and a deep drawer. Indeed, that footprint will be too big even for these storage methods. Its wooden box will need to be a stand-alone container on the floor of your study or other room. As a last recourse, it might disappear into the back of a shed or garage, where, I presume, it will rarely be seen. You might consider donating it to an interested museum with a geological collection; at the same time, any problems regarding storage will be donated, too.

Microfossils obviously present rather different problems, although they are more easily surmounted. Tubes of various sorts (Fig. 26.2D, E), gelatine capsules and cavity slides provide diverse options. Clear glass or plastic tubes with screw ends are always useful. If there is no screw cap, then plug the open end with a bung of the right size, or cotton wool. Note that cotton wool is easily removed with a pair of tweezers and removes the possibility of damage by pushing in a (too-big) bung too hard. Gelatine capsules must be kept dry, but have the advantage that they can be written on. Cavity slides are made of plastic (formerly cardboard), have a shallow well in which a specimen(s) is stored, and are sealed with a transparent, removable cover slip. All these receptacles are inexpensive and can be purchased online.

References

Arnold, B.W. & Clark, H.L. (1934) Some additional fossil echini from Jamaica. *Memoirs of the Museum of Comparative Zoology, Harvard*, **54**: 139–156.

Donovan, S.K. & Jeffery, B. (2020). A 'new' specimen of *Woodocrinus macrodactylus* de Koninck. *The North West Geologist*, **21**: 60-72.

Donovan, S.K. & Riley, M. (2013) The importance of labels to specimens: an example from the Sedgwick Museum. *The Geological Curator*, **9**: 509–514.

CHAPTER 27

LABELLING

This is just a short chapter, but an important one. However many times you may already have heard me commending the virtues of documentation, here we go again, one more time.

Consider a specimen that we will call *Agenus aspecies*. You have just collected this fine tooth, frond, shell or whatever in the field yourself – congratulations. In your field notebook you have the relevant information about this event – the date on which it was found, your locality number, the locality, the horizon (perhaps located on a measured section), grid reference and/or GPS data. In the field the specimen was wrapped in paper for protection and sealed in a bag on which (or in which) the relevant information was also recorded. At the very least, the locality and horizon have a unique number in your notebook which is also recorded with the specimen. This number is an umbilical cord, connecting the specimen to the relative data in your notebook.

You and *A. aspecies* return home. The specimen may need some light cleaning, even if only to be washed under a stream of tap water to remove the surface dirt that is adhering to it. Put *A. aspecies* onto an old newspaper to dry, and make sure that its collecting bag and any relevant labels from the field are in close association. Now is not the time to separate *A. aspecies* from its data; there will never be a time to separate *A. aspecies* from its data.

A box for *A. aspecies* needs to be large enough for it to fit without scraping against the sides, or the lid wobbling on top of the specimen. Now that *A. aspecies* has its home, it

Figure 27.1 How museums label specimens I. Label of the holotype of the Lower Devonian crinoid *Oehlerticrinus peachi* Donovan & Fearnhead, 2017, formerly classified as *Platycrinites* sp. and only known from one specimen. This specimen is in the collections of the British Geological Survey in Keyworth, Nottinghamshire. This specimen was collected before 1850, but not described until 2017. The data provided by this label essentially define the locality and horizon of this species.

needs its furniture. Any label written in the field should go into the box. Also, with your neatest handwriting and finest nibbed pen, make a clean label on thin white card or stiff paper (Fig. 27.1). This label should record all available data of relevance from your field notes.

If you have a numbering system (you should have), then *A. aspecies* will need to be numbered. The number will obviously appear in your register of specimens; on the fresh label that you have just written; on the specimen; and, if you want to be comprehensive, on the lid of the box. There are various ways to label the specimen. If it is large and robust, say a piece of Carboniferous limestone as big as a fist, then write directly on it with a narrow-nibbed permanent marker. A smaller specimen of similar lithology could be labelled directly with a smaller pen; a slip of paper, bearing the number, could be glued to the specimen; or white paint can be applied in a narrow strip and, once dry, may be written on (a thin layer of varnish may be added). Tiny specimens in tubes can have a slip of paper bearing the number inserted; a similar slip might be glued to the tube.

Do labels need further protection, above and beyond what safeguards you apply to your specimens? If your specimens are stored in a damp and cold situation, such as in a shed in the garden, then your labels will absorb moisture and may become mouldy. If you have trouble with paper-eating pests, beware. In the Geology Museum of the University of the West Indies in Jamaica, I found some of the older labels had received very precise attention from paper-eating insects, such as silverfish, cockroaches or both. They had not so much eaten the paper as eaten the ink off the paper. Labels were entire, yet blank. With care, the data could more or less be reproduced by careful examination in low light; specimens were numbered, so similar data could be resurrected from the register. After a helpful suggestion from Dr. Hugh Owen at the British Museum (Natural History) in

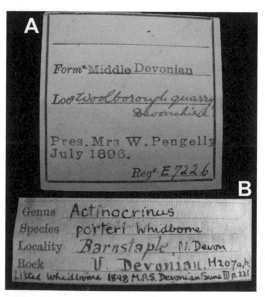

Figure 27.2 How museums label specimens II. (**A**) A Natural History Museum, London, label, for their specimen E7226. This crinoid was left in open nomenclature as camerate sp. B by Donovan & Fearnhead (2014, p. 54, text-fig. 26B, C). (**B**) Sedgwick Museum, Cambridge, specimen H207a, b. This crinoid is now classified as *Eumorphocrinus*(?) *porteri* (Whidborne, 1898). See Donovan & Fearnhead (2014, pp. 25–28, pl. 6, figs 4, 5).

London, I tried to save paper labels by wrapping them with Clingfilm. Unfortunately, this shrinks with time, wrapping paper labels about themselves, but this method should be more successful if stiff card is used instead of paper.

Do museums label their specimens this way (Figs 27.1, 27.2)? Yes, definitely: anonymous museum specimens are of little use to anyone. Part of the interest in examining a museum specimen is to find what information is stored on the labels. Museums do not dispose of old labels, so a specimen may have its original data written by the collector or the author who described it. After over a hundred years, the label and register may be the only places to discover such relevant information as locality and horizon. Remember that names of localities and formations may change with time; the label is a contemporary document and titles have changed with time. A mid-nineteenth century specimen labelled Cambrian or Silurian may be an indicator that the labeller supported the ideas of Sedgwick or Murchison, respectively (Rudwick, 1985; Speakman, 2018). Today, the same locality may be regarded as Cambrian or Ordovician or Silurian.

References

Donovan, S.K. & Fearnhead, F.E. (2014) The British Devonian Crinoidea. Part 1, Introduction and Camerata. *Monograph of the Palaeontographical Society*, **168** (643): 1–55 + i–viii.

Donovan, S.K. & Fearnhead, F.E. (2017) A Lower Devonian hexacrinitid crinoid (Camerata, Monobathrida) from south-west England. *PalZ*, **91**: 217–222.

Rudwick, M.J.S. (1985) *The Great Devonian Controversy: The Shaping of Scientific Knowledge among Gentlemanly Specialists*. University of Chicago Press, Chicago.

Speakman, C. (2018) *Adam Sedgwick: Geologist and Dalesman, 1785–1873: A biography in twelve themes*. Gritstone Writers Co-operative, Hebden Bridge, and Yorkshire Geological Society.

Whidborne, G.F. (1898) A monograph on the Devonian fauna of the south of England. Vol. III. – Part III. The fauna of the Marwood and Pilton beds of North Devon and Somerset (continued). *Monograph of the Palaeontographical Society, London*, **52** (247): 179–236.

CHAPTER 28

PHOTOGRAPHY AT HOME

Compared to the good-old, bad-old days of film, photographic paper, the darkroom (for hours at a time) and chemicals, the modern development of digital photography is extraordinarily straightforward, fast and clean. I love it, truly. Consider, for example, a specimen of the common Upper Cretaceous irregular echinoid *Micraster* sp. that you want to photograph. You plan to use your images to illustrate an article that you intend to write for the Proceedings of your local natural history or geological society (Fig. 28.1E, F). It is as big as a fist and, not conducive to a clear and well-focussed image, strongly domed and pale. A pale fossil is best photographed against a black and non-reflective background. I have a small square of black velvet which is well-suited to this task. The common direction of lighting used is from the upper left.

If you choose to produce a comprehensive suite of images of such a specimen, then I suggest six views – apical, oral (= basal), anterior, posterior, left lateral and right lateral (only the first two of these is shown in Fig. 28.1E, F). The apical view is the easy one, as the echinoid sits on the more-or-less flat oral surface. Take the photograph from directly above. Other views require stabilization. Drape the black cloth over a small, deep dish (Donovan & Lewis, 2017; Fig. 28.1A–D herein). It should now be straightforward to orientate the echinoid as required for each view. An alternative is a dish of black beads.

The problem with some white fossils, such as echinoids, is that details are not apparent: some method of increasing contrast is necessary. Adjusting for contrast and brightness in Adobe Photoshop® is commonly an improvement – I like to think of this as massaging the image. At the photography stage, a preferred method is to paint the fossil with a uniform dark colour (I prefer black), coat it uniformly with ammonium chloride (white, but matt in finish and with the black ink emphasizing subtle features such as cracks, sutures, pores and pits) and photograph the specimen in that condition. For colouring you can use Indian Ink, which is permanent. My own preference is to use food colouring, which can subsequently be washed or bleached off.

Surprising as it may seem, black food colouring is available. Paint evenly over the specimen and let it dry. The ammonium chloride needs to be contained in a puffer (Fig. 28.2). When heated it sublimates and the vapour can be puffed over the (cold) fossil specimen. In the laboratory this would be done in a fume cupboard. The best set-up

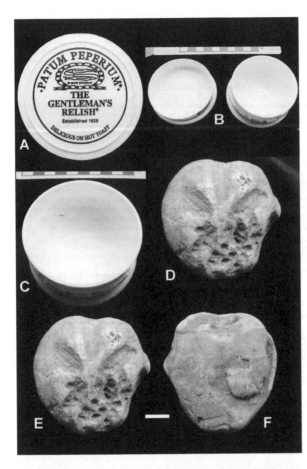

Figure 28.1 A well-organized photographic plate (after Donovan & Lewis, 2017, fig. 1). (**A–C**) Clean food container suitable for macrophotography. (**A**) Lid, width 76 mm. (**B**) Inner surfaces of lid (left) and base; scale in cm. Although we have used both of these for macrophotography, draped over with a piece of black velvet, it is the dished base (right) that provides the greatest control for orienting specimens. (**C**) Enlargement of inner surface of base; scale in cm. (**D–E**) Flint *Steinkern* (= internal mould) of *Micraster* sp. (Naturalis Biodiversity Center, Leiden, specimen RGM 792 290) photographed on the food container-base with a black velvet drape and using a Canon G11 digital camera. (**D**) Apical surface of test, mounted for photography on a square of black velvet on a food container base. (**E, F**) Apical (**E**) and oral surfaces (**F**) with the backgrounds blacked out using Adobe Photoshop®. As shown here, these images are entirely adequate for publication in a paper on echinoids. Specimen not whitened. Scale bar represents 10 mm.

Figure 28.2 Ammonium chloride puffer. The glass bulb is filled with white ammonium chloride, and it is this that is heated. The rubber tube connects to a hand puffer. Once sufficiently hot, squeezing the bulb of the puffer will eject a stream of thick, gaseous ammonium chloride for coating specimens. Scale in cm.

for this in your home is on a cooker with an overhead extractor fan. Turn the fan up high and use a gas ring to heat the bulb containing the ammonium chloride. Puffing the vapour evenly onto the fossil takes practice, but persevere.

Photograph the specimen as soon after coating as is possible, as it will sublimate, particularly in a damp atmosphere. You may have to construct your ammonium chloride

Figure 28.3 Photographic stand with a grid on the plane surface, a pair of lamps for illumination and a vertical column upon which the camera (which should be screwed into the bracket just above the level of the lamps) can be moved up and down. Apologies for the mess – I was in the process of moving house.

puffer from separate components, purchased individually – glassware, ammonium chloride, bung pierced by glass tube, rubber tube and puffer bulb, all available online.

What of steadying your camera for photography of specimens? Your specimens are small, so that any camera shake will be magnified. If you have a steady hand, the camera can be held in your hands. I prefer natural light for photography and often take my images on the windowsill in my study. There is a bookcase to my right and I steady my camera hand – or, rather, my right elbow – by resting it on a shelf. This simple set-up works for me, and many of the images of specimens in this book were photographed this way. It also has the advantage of portability. When visiting a museum to research their collections, it takes only a little ingenuity to devise an analogous set-up.

A more expensive, but certainly more elegant solution is to buy a camera stand (Fig. 28.3). The camera is held rigidly, so there is no camera shake or strange angles for photography. I use a small spirit level to ensure that the camera is truly horizontal and, thus, parallel to the base of the stand. This is obviously not portable, but it is a wise investment for anyone planning to photograph multiple specimens, perhaps in multiple orientations.

A scale bar is essential for photography. You can skirt around this if you know a standard measurement of a specimen and can say, for example, 'Width 62 mm' or whatever. However, most readers will have a little difficulty in visualizing 62 mm, whereas 10 mm is easier to work with as a comparative measure. It is better to photograph a specimen with an adjacent mm and/or cm scale in the same plane of focus (such as Fig. 28.1B, C). Then, whatever you do with an image, the specimen and a scale are linked.

If you want to prepare an image for publication, then I recommend that you read chapter 25, 'Photographic plates', *in* Donovan (2017, pp. 93–99). The one tool you will need on your computer is Adobe Photoshop® or similar program. This saves a huge amount of drudgery that was once necessary to make up a plate. Consider making a plate during the pre-digital photographic era, as I used to. To construct this, I would have printed a series of prints of each image at varying exposure times. This was time-consuming and expensive. I then spent time 'matching-up' photographs, choosing the contrast of the image of each specimen that most closely matched the others being used in the plate. The background of the plate was a sheet of photographic paper, exposed to the light and then developed to give a uniform gloss black. Each chosen image was then cut out using a sharp craft knife (careful!) under a microscope, a slow and delicate process. Images were stuck to the black background using Cow Gum rubber solution. When dry, if an image needed repositioning, it could be peeled off and again gummed in position. Scale bars were cut from white sticky labels; lettering was by dry print. This was a standard methodology, but is now a tale of an archaic ritual. In stark contrast, Photoshop® allows the user to carry out analogies of all these actions on screen: cheaper, faster, and stay out of the darkroom. Making plates of fossils is no longer a labour, but rather a pleasure.

References

Donovan, S.K. (2017) *Writing for Earth Scientists: 52 Lessons in Academic Publishing*. Wiley-Blackwell, Chichester.

Donovan, S.K. & Lewis, D.N. (2017) On the use of deep concave-dished containers in fossil invertebrate macrophotography. *Bulletin of the Mizunami Fossil Museum*, **43**: 31–33.

CHAPTER 29

DRAWING

What might you want to draw? Or what should you draw? Aside from the sketches with which you pepper your field notebook, top of your list should be neat copies of your maps of field localities and measured sections, and your notable specimens. Drawings of localities are not considered herein because, in the age of digital photography, an image is likely to be both quick and accurate. But do be sure of orientation: simply, which way was the camera pointing when you photographed the exposure – were you looking north, south or wherever?

Drawing a map

A locality map shows precisely where that prize specimen was found. For your own records, it is probably sufficient that a locality is marked and numbered on a published map of your field area, such as an Ordnance Survey (OS) map in the UK. Aim for a larger

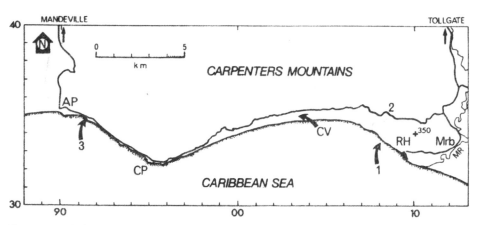

Figure 29.1 A simple locality map, suitable for a field guide (after Donovan & Miller, 1999, fig. 1). 'Locality map showing principal features of the area between Milk River Bath, parish of Clarendon, and Alligator Pond, parish of Manchester, south-central Jamaica. Key: AP = Alligator Pond; CP = Cuckold Point; CV = Canoe Valley; MR = Milk River; Mrb = Milk River Bath; RH = Round Hill; + = spot height (in m); 1, 2, 3 = Stops 1, 2, 3. Grid derived from 1:50,000 topographic sheet (metric edition) #16, "Alligator Pond".'

scale; 1:25,000 scale sheets are bigger, better and more readily available than they were in the past.

A sketch map in your field notebook will be able to show finer detail than is possible on your OS map. If the thin bed with common oysters is only exposed under a holly bush by the stream, then a sketch map can put flesh on the descriptive notes that you write. Let us suppose that you are running a field meeting for your local amateur geological society. One thing that you can do in advance is to let people know which OS sheets will be required. Take this further – in your field guide you could provide a map with the positions of your localities plainly marked. This will be an essential adjunct to your text. One way forward would be to use the OS map as a base and trace the relevant features (avoid cluttering any map or diagram with the irrelevant), numbering your localities, showing the direction of walking between them with arrows, providing a scale and north arrow, etc. (Fig. 29.1). Be sure that your caption, brimming with information as it is likely to be, includes a suitable acknowledgement such as 'Simplified from part of the OS sheet number *** *Dogger Hay and Leaky Bottom*.' See Donovan (2017, pp. 89–92) for related comments on maps.

Drawing a measured section

We have already discussed measuring sections in Chapter 5. To give a slightly different slant on measured sections, the following is adapted from Donovan (2017, pp. 85–88).

I discuss your measured sections from the position of publication; if you intend to present your data in any research publication, even something as simple as a field guide for an excursion that you are running, then you must recognize what is publishable and what is not. Your diagrams, like your text, need to be clear and neat (Fig. 29.2)

Your diagrams should be drawn to fit the space available on the page. You must plan for your figure, be it measured section or whatever, to fill a rectangular area defined by the width of a column or page in your target journal (Figs 5.2, 5.3). A measured section may be long and thin, leaving much of a page blank unless the text is divided into two or more columns. A key may take up some of this blank space, as will annotations or symbols denoting features additional to the log, such as the horizons where particular fossils occur.

An important part of clarity is the lettering of diagrams, which must be readable. Again, a hundred years ago such labelling may have been handwritten; when I was a graduate student lettering was commonly either stencilled or applied as dry print. Alternately, labels could be typed on an electric typewriter, cut out and pasted on diagrams. Today, the clean and easy way is to apply lettering in Photoshop® or a similar programme. I still draw the outline of a figure by hand, but then scan it as a .jpeg or .tiff file and label it on screen. The results are superior to any methodology that I have used hitherto.

Further, be aware of the importance of the figure caption. This is where many authors stumble. Having lavished untold effort on a course of research that leads to a diagram that is pivotal in illuminating its results, many authors then ruin it by providing an

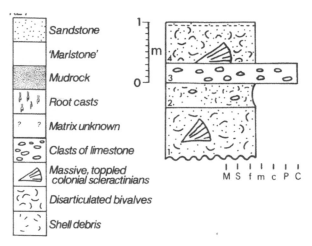

Sandstone

'Marlstone'

Mudrock

Root casts

Matrix unknown

Clasts of limestone

Massive, toppled colonial scleractinians

Disarticulated bivalves

Shell debris

M S f m c P C

Figure 29.2 A neat measured section (after Donovan & Miller, 1999, fig. 2). 'Measured section of the "raised beach" exposed in the gully above the more southeasterly fault at Stop 1C, adjacent to the section in [Donovan & Miller, 1999, fig. 3]. Key: M = mudstone; S = siltstone; f, m, c = fine-, medium-, coarse-grained sandstone; P = pebble conglomerate; C = cobble conglomerate; 1–4 = beds 1–4. Note that conglomerates in [Fig. 29.2 and Donovan & Miller, 1999, fig. 3] are matrix-supported.'

inadequate caption. Any figure can be improved by providing a truly comprehensive explanatory caption. I should have thought this obvious, but have been disabused of this innocent notion too many times. The captions of figures, tables and photographic plates are an important part of any research report. Your target as an author is to inform your readership.

Drawing a specimen

I see three possibilities for drawing specimens. If you have a talent for art, like my brother Mark (Fig. 29.3), then draw away. Yet with practice and determination, most of us improve as artists and can draw a reasonable fossil. When I taught regularly, students who bleated 'I can't draw' received short shrift. What I cannot do is defined by limitations imposed by experience, not just ability; I know that if I do not try, then I will achieve nothing. So, start drawing.

A second possibility is to trace the specimen of interest from a photographic print. A suitable computer program can be called upon to produce a worthy print of an image in the desired orientation. With a sheet of tracing paper (I prefer 90 gm/m²) taped over a print, draw around features with a fine Indian Ink pen. This should be at least 0.3 mm, as anything less than that will be so fine as to almost 'disappear'. A drawing like this has the advantage that features of interest can be emphasized more easily than on an image.

The third possibility is the expensive one, namely buying a binocular microscope with a camera lucida attachment. The microscope has a mirror attachment on the side where you customarily draw. With careful adjustment of the illumination and focus, you can see the specimen in both eyes, and the paper, pencil and your hand in one. This method of drawing resembles tracing in some ways, but from the specimen itself, not an image. Having made your pencil drawing by camera lucida, an ink copy can

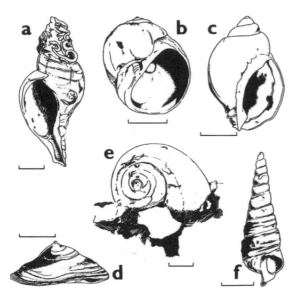

Figure 29.3 Gastropods from the Pliocene Red Crag Formation of Walton-on-the-Naze, drawn by Mark Donovan from specimens in his own collection (after Donovan & Donovan, 1989, fig. 2). All drawn in apertural view except where stated otherwise. (A) *Neptunea contraria* (Linné) encrusted by serpulid worm tubes. (B) *Natica multipunctata* S.V. Wood. (C) *Leiomesus dalei* (J. Sowerby). (D) Lateral view of the Chinaman's Hat limpet, *Calyptraea chinensis* (Linné). (E) Apical view of *Lunatia catenoides* (S.V. Wood). (F) *Potamides tricinctus* (Brocchi). All scale bars represent 10 mm.

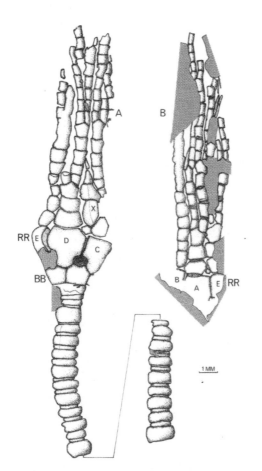

Figure 29.4 Camera lucida drawing of *Westheadocrinus girvanensis* Donovan (after Donovan, 1989, text-fig. 27), holotype, the Natural History Museum, London (prefix BMNH); drawn from part and counterpart latex casts taken from the natural moulds. (A) BMNH E49969a, crown and proxistele. (B) BMNH E49969b, reverse side of crown. Radials (RR), basals (BB), Carpenter rays (A to E) and anal X all indicated.

then be produced on tracing paper (Fig. 29.4).The camera lucida is expensive and most commonly a tool for the professional palaeontologist. However, if you have a good contact(s) at a local university or museum (Chapter 39), they may have access to a camera lucida that you could use on your next visit.

References

Donovan, S.K. (1989) Pelmatozoan columnals from the Ordovician of the British Isles. Part 2. *Monographs of the Palaeontographical Society, London*, **142** (no. 580): 69–114.

Donovan, S.K. (2017) *Writing for Earth Scientists: 52 Lessons in Academic Publishing*. Wiley-Blackwell, Chichester.

Donovan, M. & Donovan, S.K. (1989) Site wise: Walton-on-the-Naze. *Fossil Forum*, **3**: 55–61.

Donovan, S.K. & Miller, D.J. (1999) Report of a field meeting to south-central Jamaica, 23rd May, 1998. *Journal of the Geological Society of Jamaica*, **33** (for 1998): 31–41.

CHAPTER 30

SPECIALIZING IN YOUR FAVOURITE FOSSIL GROUP

When I collected my first fossils in the mid-1970s, I was wholly unselective. I filled my collecting bags with anything that came my way. This was a fine start for a budding palaeontologist – I was learning something from every specimen. Early collecting sites, both easy to reach by train from my native London, were Copt Point at Folkestone (Cretaceous Gault Clay Formation) and Walton-on-the-Naze (Pliocene Red Crag Formation). Both are very fossiliferous. With my newly-won collections I curled up with *British Mesozoic Fossils* and *British Cainozoic Fossils* (Chapter 12), and learnt to put names to fossil 'faces'.

Soon there comes to some, perhaps most of us, a desire to specialize. Part of this desire is driven by considerations of space. It is a straightforward task to accumulate a huge number of fossils once we discover suitably productive localities. This productivity, which at first is a blessing, soon deteriorates into a curse. More and more fossils take more and more space. I have worked in a few museums, so I have always had somewhere to store specimens, providing I am willing to relinquish ownership (I am). My brother Mark, an amateur, is highly selective; he has a small collection limited to dozens of well-preserved specimens, all trilobites and decapod crustaceans (crabs).

Search pattern will prejudice your collecting. If you are looking for trilobites, then you will be attuned to them, preferably complete specimens, but also fragments such as the cephalon and pygidium. My brother and I operate a simple exchange system when in the field together: if he finds a fossil echinoderm then it is mine and if I find a fossil arthropod then it is his. This almost invariably works in my favour, because echinoderm fragments are the commoner at many, perhaps most sites that we visit together. But because I collect with my eyes only about 20 cm above the rock surface, I can be relied upon to spot disarticulated arthropods. It is good to have a co-worker who has a different interest to you.

How do you further encourage your own specialization? One way is obviously to read around your subject. If you are particularly interested in a fossil group within the British Isles, then there may be one or more parts in the serial *Monographs of the Palaeontographical Society* that is of relevance (www.palaeosoc.org). (Some of the older monographs are available free online in the Biodiversity Heritage Library.) If your chosen group is invertebrate, then fill your shelf with relevant parts of the *Treatise of Invertebrate Paleontology*, published by the University of Kansas. Some parts are a little long in the tooth, but many are in the process of revision. For example, Crinoidea in three parts appeared over 40 years ago (Moore & Teichert, 1978), but the third of these, on the post-Palaeozoic species, was the first to be revised (Hess & Messing, 2011).

Whatever your desire – vertebrate, invertebrate, plant, microfossil – there is an abundance of solid information available in standard textbooks, such as Clarkson (1998), to give one example among many, and online. For the latter it is necessary to box clever. Consider, for example, the journal *Palaeontology* published by the Palaeontological Association. Bless the Association – there is a free *Palaeontology* Archive online (www. palass.org). This contains all issues apart from the past five years. To get access to the

Figure 30.1 Specialist collecting; Mississippian crinoids from Salthill Quarry, Clitheroe, Lancashire (Kabrna, 2011, locality 4), a selection of what was collected on the morning of 11th July, 2014, by Donovan *et al.* (2014, after fig. 4 therein). All specimens deposited in the Naturalis Biodiversity Center, Leiden (prefix RGM). (**A**) RGM 791 721, *Bystrowicrinus* (col.) *westheadi* Donovan, a giant crinoid pluricolumnal, showing the articular facet. (**B**) RGM 791 722, pluricolumnal with broad radice scar (upper centre) extending across several columnals and formerly part of the attachment structure ('root'). (**C**) RGM 791 723, theca of a platycrinitid monobathrid (=monocyclic) camerate crinoid with basals and high radial plates apparent. Specimen crushed. (**D**) RGM 791 724, the conical cup of the small disparid (=monocyclic) crinoid *Synbathocrinus* sp. (**E**) RGM 791 725, a small monobathrid camerate theca, probably *Amphoracrinus* sp. (**F**) RGM 791 726, a pluricolumnal encrusted by the coral *Sutherlandia parasitica* (Phillips). The crinoid is presumed to have been alive during infestation as the coral essentially engulfs the pluricolumnal through almost 360°, but there is no reaction by the host. (**G**) RGM 791 727, a strongly swollen pluricolumnal in which the living crinoid has overgrown the encrusting coral *Cladochonus* sp. (**H**) RGM 791 728, the monobathrid camerate *Amphoracrinus* sp.; compare with the somewhat smaller specimen in (**E**). Scale bar represents 10 mm.

more recent issues, you either have to be a member or you must pay for the PDF. A third possibility is to email the corresponding author, explain your interest in the paper and, politely, ask for a PDF of the paper. Most authors are delighted to provide a PDF and, if you are lucky, copies of other recent papers as well.

The title of this chapter focuses on a fossil group, but there is also a case to be made for focusing on a particular stratigraphic interval or geographic region. Retirement looms near for me, and it might be asked what I intend to research in my dotage? I shall be living in north-west England and any new fieldwork will focus on the local Mississippian (Lower Carboniferous) successions, in the White Peak of Derbyshire (Chapter 51), around the Clitheroe district of Lancashire (Chapter 50) and, as clasts, in the glacial erratics of the Fylde coast (Chapter 48). For choice, I shall collect crinoids (Fig. 30.1), echinoids and borings, but also look forward to being distracted by the unexpected.

What are the attractions of specialization? Most obviously, you will become an expert in a particular field. My late friend Stanley Westhead (1910–1986) collected in the quarries of the Clitheroe district for over 50 years, specializing in the echinoderms (Donovan, 2012). Stanley was well known to both amateurs and professionals; he had several species named in his honour; and his magnificent collection is now saved for the nation in the Natural History Museum in London. Another facet of specialization is to write and publish your observations. My late friend, Joe Collins (1927–2019), wrote over 100 papers and monographs on fossil crabs and barnacles, and was still actively researching at the time of his death (Donovan & Mellish, 2020). Joe had co-workers on both sides of the Atlantic and researched on collections from many parts of the globe. And both Stanley and Joe were amateurs. That is, neither derived a living from geological pursuits, but they were nonetheless 'professional' in respect to their collecting and research on fossils.

References

Clarkson, E.N.K. (1998) *Invertebrate Palaeontology and Evolution*. Fourth edition. Blackwell Science, Oxford.

Donovan, S.K. (2012) Stanley Westhead and the Lower Carboniferous (Mississippian) crinoids of the Clitheroe area, Lancashire. *Proceedings of the Yorkshire Geological Society*, **59**: 15–20.

Donovan, S.K., Kabrna, P. & Donovan, P.H. (2014) Salthill Quarry: a resource being revitalized. *Deposits*, **40**: 32—33.

Donovan, S.K. & Mellish, C.J.T. (2020) Mr Joseph Stephen Henry (Joe) Collins, 1927–2019. *Bulletin of the Mizunami Fossil Museum*, **46**: 103-114.

Hess, H., & Messing, C. G. (2011) *Treatise on Invertebrate Paleontology, Part T, Echinodermata 2, revised, Crinoidea, Volume 3*. University of Kansas Paleontological Institute, Lawrence.

Kabrna, P. (2011) Excursion 5. Pendle Hill and Clitheroe. *In*: Kabrna, P. (ed.) *Carboniferous Geology: Bowland Fells to Pendle Hill*. Craven and Pendle Geological Society, Lancashire, 157–165.

Moore, R.C. & Teichert, C. (eds) (1978) *Treatise on Invertebrate Paleontology, Part T, Echinodermata 2, Crinoidea*. In three volumes. Geological Society of America and University of Kansas, Boulder and Lawrence.

CHAPTER 31

WRITING DESCRIPTIONS

Describing a fossil is part of the bread and butter of palaeontology. At the very least we describe our fossils, not just if they are new species, but also specimens that show unusual features such as growth deformities. Indeed, we describe 'typical' specimens of species that may have been described many times before, but whose details are essential reference points to any discussion that we may be promoting. In describing a specimen or species, we are examining it in the minutest detail; we are observing it as accurately as we can and, in so doing, we are likely to notice certain features for the first time. Description is a necessary outcome to our observations.

When describing a specimen or specimens, your text should be supported by relevant photographs, diagrams and measurements. But what does this exercise do for you? Describing any specimen(s) is a way by which to ensure that you have examined it in the detail it deserves. It is possible to look at any specimen and appreciate its obvious features, but writing a detailed description ensures that you notice every intricacy of its form. It helps to train your powers of observation and ensure that you are not just seeing from memory. The latter is always a danger. Memory can be misleading. One of my undergraduates once described to me and the rest of his class the details of the eyes of a trilobite. It was a blind trinucleid; it had no eyes.

The specialist terminology associated with a particular group of fossils may be dense and difficult, but we can also use everyday terms to clarify our text for ourselves and

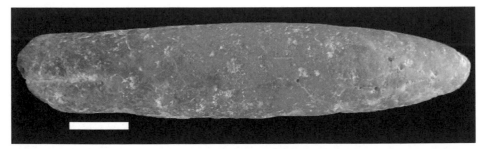

Figure 31.1 Belemnite *Belemnitella* sp., collected from float on beach between Overstrand and Cromer, north Norfolk coast; Upper Cretaceous, probably Maastrichtian. Naturalis Biodiversity Centre, Leiden, specimen RGM.1332385. Specimen uncoated. Scale bar represents 10 mm.

our readers. For example, consider the belemnite guard in Figure 31.1, *Belemnitella* sp., collected from a north Norfolk beach. It is not a fine, museum-standard display specimen, but a battered, reworked belemnite guard that any of us might pick up from the beach. How might I describe such a so-so specimen?

> Guard battered, but retaining essential features. Length c.83 mm, but incomplete; maximum diameter c.16 mm. Dull brown in colour. Circular in section, cylindrical, apex rounded; specimen markedly tapered towards apex, but not very tapered towards phragmocone. Ventral groove at abapical end straight, c.15 mm long, but incomplete. Alveolus conical, rounded, filled with chalk. Guard formed of radial calcite crystals and bored by clionaid sponges (= the trace fossil *Entobia* isp.), preserved as large chambers filled by chalk. Other traces include stellate scars; small round holes uncertainly bored, may be mechanical or chemical damage. Other surface markings due to abrasion or corrosion (= corrasion).

The equipment to write this description was suitably low technology, namely a notebook, pencil, ruler, hand lens and the relevant section of Clarkson (1998, pp. 254, fig. 8.32a–g). You may have easy access to most, perhaps all equipment on this list. To become adept at describing specimens, you just need to practise, practise and practise.

You might say, with some justification, that a belemnite has a simple shape and, in consequence, is a relatively easy specimen to describe. I would argue from experience that a simple shape can be more of a challenge to describe than a complicated shape. So, keep your Clarkson (1998) or similar texts close to hand and consider the following description taken from a published paper (Figs 31.2–31.4) *Dimorphoconus granulatus* Donovan & Paul, 1985, is notable in being as-yet unclassified in a higher taxon.

> The specimen, SM A66853a, b, consists of part and counterpart natural external moulds of a number of associated granular spines [Figs 31.2A, D, 31.3]. The spines are of two main types and the specimen preserves traces of their original organization. About 65 spines covered the animal, which was approximately 10 mm long by 8 mm wide. One counterpart [Fig. 31.2D, 31.3A] shows predominantly the tips of the spines, the other their bases [Fig. 31.2A, 31.3B]. From this we deduce that cones covered only one side of the body, which is assumed to be the dorsal surface. Some of the spines curve slightly, while others show a preferred orientation as now preserved. Both types of spine point away from a small group of five modified blunt spines at one end. We interpret the cluster of modified spines as indicating the head region, thus the curved and straight spines were orientated upwards and backwards. On the above arguments we suggest dorso-ventral and anterio-posterior orientations.

> All the spines are circular in cross-section, with finely granular surfaces [Fig. 31.2B, C]; nevertheless two principal types can be distinguished, which we term elongate and conical spines. The elongate spines, which reach six times as long as wide (2.4 mm by 0.4 mm) occur in a row of 8 or 9 on each side of the body [Figs 31.2C, D, 31.3A]

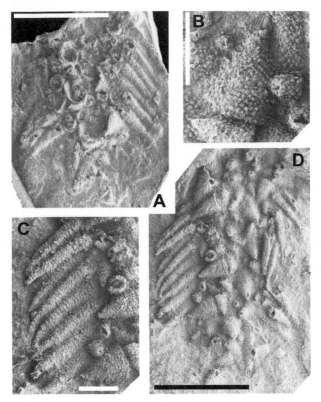

Figure 31.2 Scanning electron micrographs of latex casts of *Dimorphoconus granulatus* Donovan & Paul, 1985 (after Donovan & Paul, 1985, fig. 1), holotype, University of Cambridge Sedgwick Museum SM A66853, from the Shineton Shales (Lower Ordovician, Tremadoc) of Sheinton, Shropshire (Dean, 1968, pp. 7–8, fig. 3). **A**. General view of one counterpart showing mainly the bases of spines. SM A66853b. **B-D**. SM A66853a. **B**. Details of a large conical spine ('o' in Fig. 31.3A, B) to show granular sculpture. **C**. Detail of the left row of elongate spines. **D**. General view of counterpart showing many tips of spines. Latex casts coated with gold/ palladium. Scale bars represent 5 mm (A, D) or 1 mm (B, C).

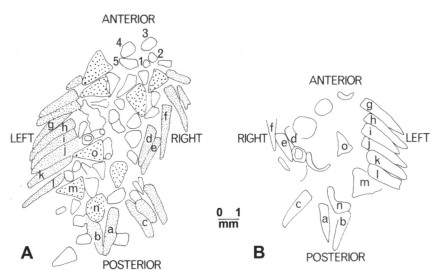

Figure 31.3 *Dimorphoconus granulatus* Donovan & Paul, 1985. **A, B**. Camera lucida drawings of latex casts of holotype, SM A66853a, b, respectively (after Donovan & Paul, 1985, fig. 2). Elongate spines shown by fine stipple; larger conical spines by coarse stipple; other spines white; 1–5 blunt spines of head region; a–o are spines which can be matched by superimposing the two figures. **A**. Dorsal view. **B**. Ventral view.

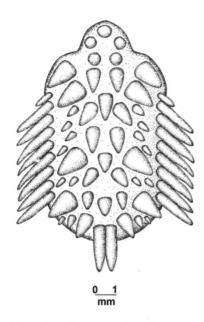

0 1
mm

Figure 31.4 Reconstruction of inferred living appearance. After Donovan & Paul (1985, fig. 3).

and the most anterior elongate spine is shortest. There seems to have been another pair of elongate spines posteriorly [Fig. 31.2D, 31.3]. Medially, the spines reach only just over twice as long as wide (up to 0.9 mm by 0.4 mm) and these conical spines occur in two sizes. There appear to have been two dorso-lateral rows each with four large conical spines (heavy stipple in [Fig. 31.3A]), separated from each other and from the lateral rows of elongate spines by series of smaller conical spines, reaching only 0.5 mm by 0.2 mm diameter at most. A group of five small modified spines, with rounded rather than pointed tips, occurs anteriorly beyond the rows of main spines [Fig. 31.2D; numbered 1–5 in Fig. 31.3A]. The other counterpart [Figs 31.2A, 31.3B] shows mainly the bases of the spines, and many can be matched between the two counterparts (lettered a–o in [Fig. 31.3]).

The base of the spines appears as smooth, shallow depressions surrounded by a narrow rim with the same granular ornament as the external surface. Although now dissolved away, the fact that no spine appears to have been crushed suggests to us that they were solid in life. The granulations on some of the lateral elongate spines are arranged in circles that resemble weak growth lines [Fig. 31.2C]. Although there is no other evidence for their mode of growth, unless the spines erupted like teeth they could only have increased in size at their bases. [Figure 31.4] is our interpretation of the appearance of a live *Dimorphoconus* (after Donovan & Paul, 1985, p. 87).

In describing a member of an unknown phylum, Donovan & Paul (1985) had no pre-existing terminology of the *Dimorphoconus* skeleton; we invented it as the description was written. We used straightforward words and language to describe our new species: if I say so myself, it still works over 30 years later.

In describing any fossil(s), my aim is to communicate, and I prefer to use language that can be understood without recourse to the dictionary. But there are technical terms that will need to be used in descriptions, and these come from two sources. First are common technical terms that you can glean from Clarkson (1998) or a similar text. Second are the specialist terms that are used by the experts on a given group and can only be found, and defined, by reference to the specialist literature. One of the most useful sources is the multi-volume work, the *Treatise on Invertebrate Paleontology*, published by the University of Kansas (paleo.ku.edu/treatise/). There are parts devoted to the groups in which I specialize, such as crinoids (in three volumes) and invertebrate trace fossils, and I keep these volumes close to hand. If you decide to specialize on a particular group, sooner or later you, too, will find that you need the relevant part of the *Treatise*. Although some parts are now long in the tooth, all are full of meat. Further, new parts are being written and chapters are published online as they become available.

Now you have written your description, what might you do with it? See Chapters 41 to 43 for comments on how to publish your palaeontological writings.

References

Clarkson, E.N.K. (1998) *Invertebrate Palaeontology and Evolution*. Fourth edition. Blackwell Science, Oxford.

Dean, W.T. (1968) *Geological Itineraries in South Shropshire*. Revised edition. *Geologists' Association Guides*, **27**: 1–48.

Donovan, S.K. & Paul, C.R.C. (1985) A new possible armoured worm from the Tremadoc of Sheinton, Shropshire. *Proceedings of the Geologists' Association*, **96**: 87–91.

CHAPTER 32

CASTING FROM NATURAL MOULDS

In the fossil record of the British Isles, mouldic preservation of invertebrates is common in many Lower Palaeozoic and Devonian deposits, amongst others. In some ways, the quality of specimens preserved as a mould can be superior to that of specimens that are still calcitic. A mould cracked from the rock may be better preserved than a calcitic shell that has been exposed to the elements. If the rock is well cemented, then the mould is likely to be robust enough to enable multiple casts to be made and distributed to co-workers. When I was a research student, latex casts could be mounted for scanning electron microscopy, which was more convenient than scanning the original specimen, which was commonly too big for the machine.

The problem with an external mould is that it is an inversion of the parent specimen. That is, it represents an 'inside out' view of the fossil. Think of a jelly and the jelly mould from which it emerged. The mould and the jelly are regular in shape and symmetry, so one can be imagined from the other. But it may be more difficult to imagine a fossil from a natural external mould, which may be preserved in a non-standard orientation of a complicated natural skeleton. You can examine a natural external mould and get a good idea of the form of the fossil. You can make a cast and get a precise impression of the fossil's form (Fig. 32.1).

First we need to go shopping (Fig. 32.2). I presume you already have an external mould that needs to be cast. So, supply yourself with:

- Bottle of liquid latex rubber (buy from an art shop)
- Dark acrylic paint such as black or red (also from the art shop)
- Sharpened pencils and a sharp pin
- A small glass jar with a tight lid
- Old newspaper (protects surfaces)
- Old clothes that can be spoiled (latex ruins fabrics – it cannot be washed off)
- Running water, preferably a tap and washbasin (sink)

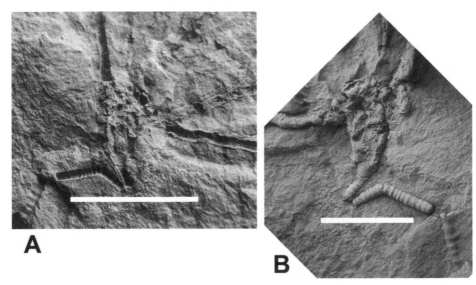

Figure 32.1 Comparison of a natural external mould and its latex cast. The Lower Silurian crinoid *Euptychocrinus longipinnulus* Fearnhead *et al.*, 2020, National Museum of Wales 2019.27G.1, holotype, details of the proximal column and lower part of crown (after Fearnhead *et al.*, 2020, fig. 3). (**A**) Natural mould, uncoated. Scale bar represents 10 mm. (**B**) Cast in red latex, coated with ammonium chloride. Scale bar represents 5 mm. Also see Figure 23.3.

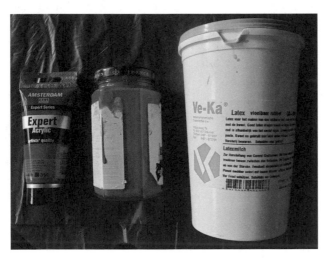

Figure 32.2 A shopping list. From left to right, tube of acrylic red paint; a jam jar of latex mixed with red acrylic paint by the author, used for casting specimens (such as Fig. 32.1B); and a pot of white liquid latex.

I commonly latex fossils on top of the washing machine: this is next to the sink in the bathroom. The best time to start is when the rest of the family is out of the house – why not send them to the movies?

Take the selected mould and make it wet all over under the warm tap; blow off the excess water, but keep the specimen damp. In the glass jar, you will already have mixed a little acrylic paint with somewhat more latex (say, about 1:20 to 1:30). Stir the paint and latex together; the colour will be darker once it has dried. Put the lid on the jar and shake it vigorously. Perhaps this should be done the day before, to give a chance for the inevitable bubbles in the latex to burst.

Always be sure to screw the lids back on the latex bottle and your jar of coloured latex. Invert them, so that all the surfaces of the bottle and jar are coated with the liquid, and keep them somewhere cool and dark; I put mine in a cardboard box in a wardrobe. Every time that you cast a specimen, put the lid back on the jar and bottle – tight – go through the inversion ritual and return to that cool, dark place. This will ensure that you get maximum usage out of a substance that can set hard overnight if not treated with respect.

Some people choose to paint the latex on a specimen with a fine artist's paintbrush. This may generate a lot of bubbles, which will need bursting with a sharp pin, and the brush is soon ruined, set solid with latex. Rather, I prefer to drop the coloured latex onto the damp fossil using your sharp pencil, and blow it off. Be sure to leave a thin film of latex on the damp fossil. If there are any bubbles in the latex, burst them with a pointed object such as a pin or a well-sharpened hard pencil (at least 2H). Leave the latex to dry under a gentle heat, such as on a windowsill, but perhaps out of direct sunlight.

Once the latex is dry, repeat the exercise. While there should not be any need to moisten the latexed surface, I find it does help on some specimens. Again, blow the layer thin and leave to dry.

The third and subsequent layers can be thicker. Coat and dry over a period of days. Keep watch for bubbles in each new application and burst them. Eventually, multiple layers of the latex will build up into a thick coating of, say, 5 mm or so. Once your final layer is dry, fold up a corner of the latex and gently peel it away from the specimen. The results can be exquisite (Fig. 32.1B). Protect your latex casts from the air and strong light, which will harden them; keep them in a small box with a tight lid or they will soon be covered by particles of dust and hair.

It is possible to make artificial external moulds for casting. A shell in a well-lithified sandstone or mudrock can be immersed in a bath of 10% hydrochloric acid. This will dissolve out the calcite of the skeleton while leaving a pristine external mould for casting.

I have emphasized the casting of external moulds. Such a latex cast will show the external surface of the specimen. A latex cast of an internal mould will probably be less useful, showing as it does the features of the internal surfaces of the shell, which are rarely of use in the detailed identification or description of a specimen.

Why colour the latex? If used uncoloured, the cast will be translucent. Recognizing detail on such a cast will be difficult unless it is coated with ammonium chloride (Chapter 28). Even if coloured, the detail of a latex cast can be enhanced with a layer of ammonium chloride. Ammonium chloride can be removed simply by washing with tap water or, if left, will eventually sublimate, particularly in a damp atmosphere.

My late colleague, Joe Collins, would describe and latex new species of fossil crabs. He would then make plaster casts from these moulds. The artificial casts would go into

his own collection and the original specimen would be donated to the Natural History Museum in London.

I also make latex moulds of borings, modern specimens off the beach (Fig. 32.3) and those fossil specimens that have not been infilled. Such moulds can expose the fine detail of a boring without the necessity of breaking into the specimen.

References

Donovan, S.K. (2013) A Recent example of the boring *Gastrochaenolites lapidicus* Kelly and Bromley and its producing organism in north Norfolk, eastern England. *Bulletin of the Mizunami Fossil Museum*, **39**: 69–71.

Fearnhead, F.E., Donovan, S.K., Botting, J.P. & Muir, L.A. (2020). A Lower Silurian (Llandovery) diplobathrid crinoid (Camerata) from mid-Wales. *Geological Magazine*, **157**: 1176-1180.

Figure 32.3 A modern boring moulded in latex. Club-shaped *Gastrochaenolites lapidicus* Kelly & Bromley, from the Recent of north Norfolk (see also Donovan, 2013, fig.2A), Naturalis Biodiversity Center, Leiden, specimen RGM 780 601. The 'cracks' in this specimen show where the specimen was glued back together. Scale bar represents 10 mm.

CHAPTER 33

PROBLEMS WITH PRESERVATION

In this chapter, when I discuss problems, I am thinking of what happens to some specimens after they have been collected. You bring them home and give them pride of place in your collection. You may not look at said specimen for some months or years, and when you do … something nasty has happened. There are three common 'nasty' scenarios that I want to discuss: sea salt crystallization; drying mudrocks; and pyrite disease.

Sea salt crystallization

My personal experience of salt water damaging fossils is with echinoids from the Chalk. At Margate in Kent the large holasteroid *Echinocorys scutata* Leske can be collected from the Chalk cliffs, where they occur in an echinoid-rich band, or may be found washed up on the beach. On the north Norfolk coast between Overstrand and Cromer (Chapter 47), I beachcomb for *Echinocorys*, which is most likely derived from offshore. From both sites I can collect fine specimens that subsequently, perhaps over a period of years, start to disintegrate, whereas others remain unaffected.

The specimens that disintegrated were collected wet from seawater or salt spray. As they dried out, salt crystals grew in pore spaces within the Chalk infilling the test. I presume that the crystals induce stress in the Chalk that may act to wedge the test apart, a piece at a time. If the test is stored in a place that is intermittently damp, the salt is likely to dissolve and recrystallize, perhaps many times.

In an attempt to remove the salt, I collect *Echinocorys* and soak it in a bowl of tap water overnight. The next day I drain off the water, and wash the echinoid and the bowl. I then repeat the soaking cycle with fresh tap water. Repeating this process over a few days does seem to help. Certainly, I have not had an 'exploding' Chalk *Echinocorys* for many years. The recommendations of Larkin (2019; see 'Pyrite disease', below) for keeping specimens dry might also be followed.

Drying mudrocks

Mudrocks vary all the way from the well-lithified to the friable. It is at the latter end of the spectrum that problems occur. A friable mudrock will gradually fall apart as it dries out (Anonymous, 1998). This may be an advantage, as it may release a prized fossil, but the fossil itself may be part of the disintegration. This process can be slowed down, in the short term, by keeping the mudrock damp. I have also seen fossils in mudrock clasts that have been embedded in plaster, which has gone some way to stabilize the whole specimen.

Converse (1987, p. 6) recommended stabilization by using a water-based glue. 'When specimens start cracking during the drying process ... a water-based glue MUST be used. A water soluble form of this glue is currently sold in department stores, construction suppliers and hardware stores ... This polyvinyl acetate material is thinned TO THE VISCOSITY OF MILK using water' [author's original emphasis]. I always have PVA glues to hand for repairing limestone fossils that were accidentally broken into two or more fragments when collected. Converse recommends the fossil is immersed in his milky solution for 2–3 days, then removed and dried slowly.

Pyrite disease

In writing this section I have been greatly helped by a recent set of thematic papers published in the *Geological Curator* (volume 11, no. 1, June 2019, pp. 1–68) and entitled 'Pyrite oxidation: where are we now?' Pyrite disease, pyrite decay or pyrite rot (all the same thing) is a potential affliction of any fossil more or less preserved in this mineral. Pyrite is an iron disulphide, FeS_2, which forms under anaerobic conditions and breaks down in an oxidizing (oxygen-rich) environment. See Larkin *et al.* (2019) for a list of formations, sites and fossils prone to pyrite disease.

In some circumstances pyrite may be beneficial, such as in preserving 'soft' (unmineralized) tissues and helping to resist compaction. Fossils may be pyritized in one of several ways (Miles, 2019):

- Permineralization, where pyrite is precipitated in pore spaces of, for example, bone.
- Replacement (or pyritization) when calcium carbonate is replaced by pyrite, such as the aragonite of ammonite shells.
- Mineral casts and moulds, when void space is infilled after a shell has been dissolved away.
- Coatings and linings on the outsides or insides of fossils.

Pyrite reacts with oxygen and water; these are the enemies of your pyritized fossils. The products of pyrite disease (= breakdown) are commonly sulphuric acid, sulphur dioxide and hydrated iron II (ferrous) sulphates (Miles, 2019). It acts fastest under conditions of high temperature, high humidity and with fine-grained pyrite. Inappropriately high temperatures may promote the start of pyrite disease in days. Signs of decay include:

- Loss of surface shine.

- Distinctive sulphurous smells.

- Pale, powdery efflorescence, with a 'cigarette-ash like appearance' (Miles, 2019, p. 6).

- Expansion cracks.

- Acid burns on drawers, boxes and labels.

When I first started to collect pyritized ammonites from the Gault Clay Formation in the mid-1970s, received wisdom was to soak them in Savlon to kill bacteria, which were thought to be involved in the process (now shown to be wrong) and to paint them in varnish (which did not help much, at best). Today, it is recognized that oxygen and humidity need to be reduced (Fenlon & Petrera, 2019; Larkin, 2019). The most effective treatments are expensive, probably more than might be expected except in a major museum (Allington-Jones, 2019). 'A cheaper and easier alternative for many specimens is to use a suitable lidded polyethylene or polypropylene container (they are usually impervious to moisture and oxygen), with a desiccant [such as silica gel] placed along with the specimens' (Larkin, 2019, p. 34). I recommend that you read Larkin's paper if pyrite fossils are a major part of your collection.

References

Allington-Jones, L. (2019) Counting the cost (of low-oxygen storage projects). *Geological Curator*, **11**, 27–31.

Anonymous. (1998) Storage concerns for geological collections. *Conserve O Gram*, **11** (2): 1–4.

Converse, H.H., Jr. (1987) Preparator's techniques. *The Ecphora*, **3** (3): 6–7.

Fenlon, A. & Petrera, L. (2019) Pyrite oxidation: a history of treatments at the Natural History Museum, London. *Geological Curator*, **11**, 9–18.

Larkin, N.R. (2019) Keep calm and call the conservator: it is only pyrite decay and your specimen may be salvageable. *Geological Curator*, **11**, 33–38.

Larkin, N.R., Buttler, C.J. & Miles, K. (2019) UK and Irish locations from which geological or palaeontological specimens are known to be prone to pyrite oxidation. *Geological Curator*, **11**, 19–26.

Miles, K. (2019) The role of pyrite in fossilization and its potential for instability. *Geological Curator*, **11**: 3–8.

THE WIDER FIELD: GETTING INVOLVED

CHAPTER 34

COLLABORATION

Sometimes I enjoy fieldwork on my own, and sometimes it is best with one or more other people. Sometimes I enjoy researching on my own and at other times I am part of a collaboration, pursuing a research project. All of these possibilities are fine, producing satisfying results.

Collaboration is not for everyone. Some collectors and researchers prefer their own company. One of my scientific 'heroes' is the late Charles Taylor Trechmann (1884–1964). Trechmann was independently wealthy, lucky man, and was devoted to his scientific research. In the summer, he worked on the palaeontology and archaeology of the Permian and Quaternary of north-east England (he lived in Co. Durham). When winter came on, Trechmann booked a ship passage to the Caribbean, where his main interests revolved around the Cretaceous and Cenozoic molluscs. He wrote many papers and monographs on these subjects, but, with one exception, all were by Trechmann alone (Trechmann & Woolacott, 1919; Donovan, 2001). Even when Trechmann collected new non-molluscan taxa, he sent the specimens to recognized experts, but did not publish in collaboration with them.

Unlike Trechmann, I enjoy fieldwork and related activities with my research 'group', far flung and never all together in the same place. We have interests in common and interests that are different. We all have our personal stores of knowledge from which the rest can learn something new; what is new to you might be old to someone else. A few years ago, I visited a quarry with two colleagues in the Clitheroe area of Lancashire, one behind locked gates (one colleague had access to the key). I have only visited this site a few times over almost 40 years and something interesting always pops up. Our main aim was to collect Mississippian echinoids (we were successful), but one bed intrigued us. It preserved clumps of elongate fossils reminiscent of the bristles of a coarse broom. I had seen something like them before and suggested that they were particularly robust sponge spicules. This was met with some little doubt by my colleagues, but I was delighted later to receive an email from one of them confirming my identification. Indeed, old man Donovan knows some fossils!

The sort of collaboration that I am discussing herein is concerned with fieldwork and discovery, but not writing (see Donovan, 2017, pp. 105–108). It is a simple correlation – the more pairs of eyes that you can bring to bear on a problem in the field, the greater the likelihood that the right fossil(s) will come to light. There is obviously a sliding scale here, from your own pair of eyes to tens of pairs on a field excursion. My preferred size of a group of co-workers might be two to four or so. It is also constructive to embrace a range of interests. As an exotic example, on my last field excursion to Antigua (Chapter 52), there were Jim Toomey and Roger Portell (essentially catholic in their interests, but with a taste for crabs), Dave Harper (brachiopods) and myself (echinoderms, trace fossils). We were looking for our own groups, but also had our eyes open for everyone else's 'beasts'.

Of course, you must be selective in choosing with whom to work. I leave you to write your own list of reasons for excluding collector X from your field group. Being a poor collector is not necessarily a good reason to exclude someone; experience says that they may find excellent material, but just not so much of it. But someone who is secretive, who gathers their specimens to them, who never says 'These are your specialism, this bed is loaded with them', they might be better encouraged not to join your group.

A collaborative mentality encourages dissemination of knowledge. If you are a member of a local field club that is investigating a productive site, bed or formation, then what better way forward than for everyone to take on the responsibility for a particular aspect of the palaeontology or the geology? In 1948–1967, the Freelance Geological Association in south London was a group of like-minded collectors at a time when quarries in south-east England were more common than today and, certainly, more accessible. Inland sites of the Gault Clay Formation (Cretaceous) were highly productive, and some members developed a strong, specialist expertise in aspects of the fauna such as crustaceans (Joe Collins; e.g., Collins & Collins, 1959) and ammonites (Ray Milbourne; e.g., Milbourne, 1956).

What makes an ideal collaborator in the field? Enthusiasm and perseverance are essential, but not necessarily expertise, at least to start with. I have happy memories of the days when I was a nascent amateur geologist in 1975. In those days I was entirely self-taught, including an O-level in geology in 1976. But fieldwork was (and still is) the thing. My guide was Arkell et al. (1954), a book that, more than 20 years after it was published, took me to a mixture of exciting collecting sites and infilled quarries. It was a bit of a lottery, but I appreciated that times had changed. The 72 authors of Arkell et al. might be considered my first collaborators, some of them already long dead when I bought the book from (I think) the shop of the late lamented Geological Museum in London.

But I like to think that my new enthusiasm for fossils was contagious. In those days, until I went to university in September 1977, I was a telephone exchange maintenance engineer working for the Post Office (now British Telecom). Several of my fellow engineers were 'led astray' by me and, I trust, enjoyed the experience. We visited coastal exposures (Folkestone, Walton-on-the-Naze), and disused chalk quarries in north Kent and Surrey. My younger brother, Mark, could be lured anywhere that promised either trilobites or crabs; years later, we actually did find a *Notopocorystes* in the Gault Clay Formation at Folkestone. My friend Paul Vidler, who actually studied geology at school, shared a day with me around an Aylesbury that disappointed with infilled quarries, but later repaid our perseverance when we found ammonites built into the walls of (I think) Waddesdon Manor (Donovan, in press). All are treasured memories of happy and educational times.

References

Arkell, W.J. & 71 others. (1954) *Directory of British Fossiliferous Localities*. Palaeontographical Society, London.

Collins, J.S.H, & Collins, J. (1959) Recognition of fossils part 1. An introduction to the decapod and cirripede Crustacea of the Gault. *Freelance Geological Association Journal*, **2** (1): 8–15.

Donovan, S.K. (2001) The publications of Charles Taylor Trechmann (1885–1964), a notable amateur geologist from the northeast of England. *The Vasculum*, **86** (3): 21–25.

Donovan, S.K. (2017) *Writing for Earth Scientists: 52 Lessons in Academic Publishing*. Wiley-Blackwell, Chichester.

Donovan, S.K. (in press) The Amateur Geologist. *Geology Today*.

Milbourne, R.A. (1956) The Gault at Greatness Lane, Sevenoaks, Kent. *Proceedings of the Geologists' Association*, **66** (for 1955): 235–242.

Trechmann, C.T. & Woolacott, D. (1919) On the highest Coal-Measures or 'Zone' of *Anthrcomya phillipsi* in the Durham Coalfield. *Geological Magazine*, **56**: 203–211.

CHAPTER 35

SCIENTIFIC SOCIETIES

I write this chapter on the assumption that you want to meet other palaeontologists. There are many scientific societies and not a few have programmes of interest to you. I remember that, in the pre-www days of the mid-1970s, such societies were a mystery to me until I read Peter Cattermole's *The Amateur Geologist* (1968). This put me on the track of the Geologists' Association (GA), which I joined about 1976, and of which I have been a member ever since, apart from a short hiatus (after a row about a field guide).

What is the purpose of a scientific society? Foremost, it brings like-minded people together. For example, the GA has a dense programme of meetings. There is a lecture in central London every month apart from the summer (when we should all be in the field); all members may attend and bring their guests. When I last lived in London (1998–2001), attending the monthly GA meeting was a fast way to get in touch with my co-author, Joe Collins, who had no telephone or email and who lived at the other end of town to the family Donovan. So, I could progress my joint research with Joe and enjoy the lecture by the guest speaker. Add to these talks the many field meetings (commonly one-day, but also longer) and occasional conferences, and there is little excuse not to know some fellow members with like interests.

There are many other advantages in belonging to a society, but I want to highlight two – grants and publications. Few scientific societies have a surfeit of funds, but most like to support relevant research and other aspects, such as conferences, with small grants.

Publications are needed for posterity. Science that is not published will be lost (and so much has been lost). Any society worth its salt will have a newsletter, at least, and probably also a research journal. The newsletter is there for society news, informing the membership of details of the next lecture meeting, reporting on the recent field trip, and so on. It provides a taste of what the society does. The journal is a different enterprise altogether, recording that which needs preserving, mainly research papers, but also including diverse additional articles such as book reviews, more science-rich reports of field meetings than the newsletter, the society's proceedings, and so on.

Scientific societies can be split into the big (national and international) and small (local). The big societies can be specific in their interests – geology or palaeontology, for example – whereas local societies commonly cater for a range of interests, such as natural history and archaeology. For the current readership, there are two big palaeontological societies that you might consider joining. The Palaeontological Association (UK) publishes *Palaeontology* and *Papers in Palaeontology*, as well as an excellent newsletter. There is an annual meeting just before Christmas, with talks and posters, an annual dinner and a field trip (not always the best idea in December). Other meetings include the annual 'Progressive Palaeontology' for presentations by students.

The Paleontological Society (USA) publishes *Journal of Paleontology*, *Paleobiology* and a newsletter that appears irregularly. Meetings of various sorts occur throughout the USA, but the main 'cluster' occurs as part of the Annual Meeting of the Geological Society of America. This is an agglomeration of conferences of various societies, so, for four days of talks and posters, there is always something palaeontological going on.

I have already mentioned the GA. It lies at an intersection of the big and small societies. It has regional groups and other affiliated organizations which organize meetings around the country. Future meetings of all these associated societies are listed in the glossy *Magazine of the Geologists' Association*, published four times per year, along with relevant web addresses. There is also the *Proceedings of the Geologists' Association*, a scientific research journal published six times per year and which commonly includes papers on or relevant to palaeontology. In short, like Donovan in the mid-1970s, there is much to recommend the GA as your first scientific society.

What of the small (local) societies? Hunting on the web may lead you to the one nearest and most relevant to you, or you can trawl the pages of the *GA Magazine*. There are two sorts of relevant organizations, either geological or natural history societies. Examples of the former to which I am a member include the Manchester Geological Association (MGA) and the Geological Society of Norfolk (GSN). In retirement, I shall be living in Manchester, and look forward to attending the lectures and field meetings of the MGA. I already contribute to the MGA journal, *The North West Geologist*, co-published with other local societies. I have an active field programme in north Norfolk (Chapter 47) and use the *Bulletin* of the GSN as a source of new information on a favourite field area.

As you will have gathered, I am enthusiastic in recommending membership of your local society as a way to meet and interact with fellow local experts and enthusiasts at many levels. And do not be shy of writing for your local geological journal and/or newsletter. If you don't write for it, who will?

References

Cattermole, P. (1968) *The Amateur Geologist*. Lutterworth Press, Cambridge.

CHAPTER 36

CONFERENCES

Conferences are one of the 'frills' of palaeontology, an expense of both time and money, but also an investment. There will be a registration fee. Most conferences occur distant from your home town, so there will be the time and expense of travel. If the meeting lasts more than one day, then you will need a hotel room, preferably close to the conference venue. Local travel, eating out – it all adds up. And yet you should attend as many relevant conferences as you can afford, as an investment.

For me, 'relevant' is defined by palaeontology, in general, and specifically echinoderms and trace fossils. I have attended meetings on these subjects for several reasons. The main fact is that if something new, exciting and relevant to your interests has been discovered, then you will probably hear it first at a conference – as a talk, on a poster or by word of mouth. As a general rule of progression, an idea is first aired at a conference (or conferences), it is then published as a research paper and, assuming it is perceived to be sufficiently important, eventually appears in standard textbooks. If you read something in a textbook, it may be significant, but it is not the latest hot-off-the-press breakthrough: that was years ago.

Conferences are also excellent places to meet the many experts in your field of interest. At meetings of your local scientific society, you will have many friends sharing your local interests. Going to a conference, with a varied audience of students, academics and amateurs, is the next level of interaction. You have read papers by Dr. Zed with interest – they work on the same age of fossils as you, but have field areas such as, say, Scandinavia or south-east Asia, far more exotic. Let Dr. Zed know what you have found and how you think it may (or may not) relate to the talk that they have just presented. It will be a strange academic who will not be interested.

Let me give an example. Ron Pickerill and I were both ex-University of Liverpool Ph.D. students (Ron graduated before I arrived) when we met at the bar on the first evening of the Palaeontological Association Annual Meeting in December 1989, coincidentally in Liverpool. Conversation got around to trace fossils, Ron's principal interest and, at that time, something that I no more than dabbled in (things were about to change). I lamented that Ron could not get from Canada (where he was professor at University of New Brunswick, Fredericton) to Jamaica (where I was senior lecturer at the University of the West Indies) where there were numerous, undescribed trace fossils in the Paleogene siliciclastics exposed north and east of Kingston. To cut a long story short, Ron got to Jamaica in February (just two months later!) and we published the first of many joint research papers before the end of 1990.

Talks

Listening to and presenting talks is the bread and butter of conference life. Conference talks are commonly brief – 10, 15 or 20 minutes – apart from 'key note addresses' by leading experts giving invited presentations of, say, half an hour to 45 minutes. If you sit through a day of talks you will hear a lot of information; make a note of things that you want to remember. A day of 15 minute presentations might include, say, 32 talks spaced out in chunks of eight and separated by coffee/lunch/tea breaks. Phew!

There will be an accompanying book of abstracts of, say, 200–250 words each. This is an *aide-mémoire* for the future. You can annotate the abstracts or take notes if you wish.

I rarely sit through a whole day of talks, although I did when I was younger. My attitude is that there are some talks that I will attend, in my subject areas or by my friends, and the rest is a movable feast. To be blunt, I no longer have the stamina to sit through everything, so I am selective. If you, too, recognize this truth, you may enjoy the meeting much more. Do something different – view the posters, have a coffee, cruise the trade show or chew the fat. They are all relevant parts of the conference experience.

After your first conferences, you may want to give a talk yourself – a big step in science. Some good references of giving presentations include Levin & Topping (2006), Bradbury (2006), Hall (2007) and O'Rourke (2008). See Donovan (2017, pp. 39–41) for helpful comments on writing your abstract: remember, the abstract is the only published remnant that will survive after you present a talk.

Posters

Posters are a different way of presenting information at conferences. Instead of giving a short talk of limited duration, a poster may be displayed for hours, perhaps days. This may be a good or a bad thing. It is bad when posters are just displayed like some new wallpaper. I remember one conference in the UK where the conference posters were displayed along the corridor to the toilets. This did not prove to be a position conducive to casual viewing.

The American way is better. There are sessions of posters just as there are sessions of talks. Posters are displayed in their own area, a large room or even part of an extensive exhibition hall. The presenter(s) of a poster stands with it for at least two hours, and are thus available to discuss it with the interested. So, the potential is there for you to speak much longer than any talk and to stand up for hours on end! Talks are sprints; posters are marathons. But a good poster session is a joy, giving you a chance to interact directly with the author and discuss their research in detail.

Field trips and trade shows

A good geological or palaeontological conference will have an associated field trip or, better yet, a choice of field trips. Assuming that the conference is held at a location distant from your home, then go on the field meeting and see some new (to you) fossiliferous rocks. There will probably be a specially written field guide and, if you are not far from civilization, a good pub lunch. If you are too far from a pub, then I predict that the specially provided packed lunch will include either a hard-boiled egg or a chunk of Danish Blue cheese (or both). They always do.

Really big conferences (like the Annual Meeting of the Geological Society of America) have an associated trade show. These are great. For the palaeontologist, there will be diverse publications, stalls run by palaeontological societies, field equipment, maps, exotic fossils for sale and, essentially, the works. The experienced conference attendee will also be sensitive to what is free, such as outdated copies of journals, book catalogues, pens, pencils and sweeties.

Take a shopping bag and remember to take your credit card.

References

Bradbury, A. (2006) *Successful Presentation Skills*. Third edition. Kogan Page, London.

Donovan, S.K. (2017) *Writing for Earth Scientists: 52 Lessons in Academic Publishing*. Wiley-Blackwell, Chichester.

Hall, R. (2007) *Brilliant Presentations: What the Best Presenters Know, Say and Do*. Pearson Education, Harlow.

Levin, P. & Topping, G. (2006) *Perfect Presentations!* Open University Press, Maidenhead.

O'Rourke, J. (2008) *The Truth about Confident Presenting*. Pearson Education, Harlow.

CHAPTER 37

JOURNALS AND MAGAZINES

New ideas and observations in palaeontology may first be reported at a conference, but publication is a way to preserve them for posterity. If it is not published, then it will not exist. In theory, a published scientific paper should be available, one way or another, to anyone who wants to read it (see Chapter 38). In Chapter 35, I introduced several journals (there are many, many more) published by named palaeontological and geological societies. Herein, I discuss a few other journals that should be of interest to you.

Bulletin of the Mizunami Fossil Museum

This seems exotic – I start with a Japanese palaeontological journal? I was first made aware of the *Bulletin* by my late friend, Joe Collins, who published in it regularly (Donovan & Mellish, 2020). The Bulletin, like those journals listed in Chapter 35, is peer reviewed. That is, if you submit a paper for possible publication, it will be sent to an expert in the field for assessment (Donovan, 2017, pp. 127–129). This is a form of quality control widely practised by scientific journals.

The *Bulletin* is of particular note; it is a focus for papers on fossil crabs. Joe Collins published many of his papers on fossil crabs in the *Bulletin*; Dr. Hiroaki Karasawa is a specialist on them; and the rest of the community of palaeocarcinologists (that is, workers on fossil crabs) took due note and followed Joe's lead. It is not unusual for half the papers in any issue (published annually) to be on crabs. And there is one further and highly attractive point about the *Bulletin* – it is available online for free. Indeed, it is one of the few peer-reviewed research journals that are available for free (but also see *Caribbean Journal of Earth Science*).

Proceedings of the Yorkshire Geological Society

The *Proceedings* is published twice a year and dates back to the nineteenth century. It is devoted to all aspects of the geology of the north of England. I enjoy this specialist geographical flavour. I have many interests in the palaeontology of the north of England (see, for example, Chapters 48, 50, 51) and the *Proceedings* often addresses one or more of these. Further, the website (it is published by the Geological Society Publishing House) lists papers 'in press' that we might expect to see in the next issue.

If your interests are similarly focused on the north of England, then it might be worthwhile joining the Yorkshire Geological Society. Apart from receiving two copies of the *Proceedings* per year, you will also have electronic access to every previous issue – a tremendous resource.

Bulletin of the Geological Society of Norfolk (BGSN)

I include *BGSN* as one example of the sort of fine publication produced by many of the smaller scientific societies. It is published annually; the journals of some other societies appear less regularly. Many back issues of *BGSN* are available free on-line (http://www.norfolkgeology.co.uk/bulletin.htm). Each issue includes two to four papers, relevant to the title.

This emphasizes an important constraint on academic publishing; everyone wants to publish in the 'big' journals, but fewer submit to the 'smaller' journals even if they are peer reviewed, as is *BGSN* (Donovan, 2011). An editor can only publish what is submitted, which may not be what the readership wants, but is what is available. (Believe me, I know this, as a former editor of a 'small' journal, the *Journal of the Geological Society of Jamaica*; Donovan, 2016). Journals like *BGSN* need to be appreciated for what they are, not what some may imagine they might be.

I also note that many contributors of small journals like *BGSN* keep coming back; they like the way that they are treated and the style of the publication, so sooner or later they return with a further paper. In turn, this means that a 'small' journal gets known for particular aspects of geology and palaeontology, as papers are written by the same authorship (see *Bulletin of the Mizunami Fossil Museum*, above).

Geology Today

Geology Today publishes a mix of news items and semi-popular articles on diverse aspects of geology, including palaeontology. It appears six times per year as a joint initiative by the Geologists' Association (see Chapter 35), the Geological Society of London and Wiley-Blackwell. Every issue is packed with information and beautifully produced. Its great strength is its readability and diversity. Also, it is published in colour throughout.

Deposits

Last, but definitely not least: subscribe to *Deposits*! Do it now! But times are changing. Formerly *Deposits* was published four times per year, appearing on glossy paper throughout, and included a glorious assortment of articles in a rainbow of styles. While geological in tone, palaeontology is a common theme. There are short contributions – book reviews, guest editorials and the like – with descriptions of interesting sites (many with good fossils) and articles looking in more depth at particular subjects. Of all the journals discussed in this book, *Deposits* is the one unashamedly focused on the amateur. But it has now changed to an on-line only publication, and I hope it continues to thrive in its changed environment. There is a lot available on the *Deposits* website, too.

References

Donovan, S.K. (2011) Big journals, small journals, and the two peer reviews. *Journal of Scholarly Publishing*, **42**: 523–527.

Donovan, S.K. (2016) Editing in Jamaica 1989–1998. *Publications*, **4** (10): 5 pp. doi: 10.3390/publications4020010.

Donovan, S.K. (2017) *Writing for Earth Scientists: 52 Lessons in Academic Publishing*. Wiley-Blackwell, Chichester.

Donovan, S.K. & Mellish, C.J.T. (2020) Mr Joseph Stephen Henry (Joe) Collins, 1927–2019. *Bulletin of the Mizunami Fossil Museum*, **46**: 103–114.

CHAPTER 38

OFFPRINTS, PDFS AND FILING

I have talked elsewhere of accumulating relevant books and what they might be (Chapter 12), and I have also enthused over certain journals and magazines that you will find of interest (Chapters 35, 37). I am now going to look at the way to use and accumulate the separate chapters and research papers that may be of direct relevance to your interests. For example, if you collect and have an interest in Palaeozoic arthropods, it may be convenient to have separate filing systems for relevant literature on trilobites, eurypterids, insects, the Burgess Shale and so on. There may be a lot of pertinent information in your growing accumulation of books and journals, but can you find it when you want it?

There are, broadly, two ways to organize your 'knowledge base', if you will. You can have actual physical copies, photocopies and printed PDFs of the same and/or you can have an electronic filing system, including your PDFs, in which the searchable attributes of each are listed, most importantly where it can be found. As a green research student, I collected copies of papers and recorded what I had in a card file (the 1980s equivalent of a spread sheet), but I soon trimmed down to just accumulating hard copies of every relevant paper (PDFs would not be invented for many years); maintaining a card index was just too much work.

Offprints and PDFs

Life used to be more physical. I would publish in hard copy (then the only option) and then would either receive free, or buy, or a bit of both, offprints (= separate copies) of my paper. Typically, a journal might send 50 free offprints – ample if I was sole author – but if I had, say, two co-authors, that worked out at about 17 each. For distribution amongst others with similar interests, 17 did not go far. And if you got too many offprints of any paper, you would probably have spares left over that you would need to store (Donovan, 2009, 2017, pp. 151–155).

Life is now streamlined and storage is as a PDF inside your laptop. Today, few journals publish in hard copy only, and those are mainly the organs of amateur scientific societies. When I publish an article, it is unusual to receive offprints. Rather, I get a PDF – not unexpected, as the journal is likely to be published either in hard copy and online or just online. PDFs are the offprints of the twenty-first century. Any problem of storage is solved by your hard drive or memory stick.

I have been talking about preserving my own publications as hard copies and PDFs (these days, usually both), but a similar set of comments applies to any collection of research papers. Whatever collection of papers you are amassing for your own use, your collection will need structure. There are all manner of possibilities, but I use a system as simple as possible for my trace fossil papers, housed in three drawers of a filing cabinet and in alphabetical order. In addition to A, B, C and so on, I have one or more separate folders to hold the papers of important and prolific authors, such as Bromley and Pickerill.

An electronic file of PDFs, perhaps cross-referenced to a spreadsheet, has obvious potential, but you will have to be on the ball to keep it up to date. I prefer the physical, so that I invariably have a small pile of papers that are awaiting filing. Once the pile reaches a height that pricks my conscience, I devote 20 minutes or so to filing papers. I keep no other record of what I have on file, but can generally remember the important papers. It is the forgotten obscure papers that give joy: I bump into them when searching for something else and am always delighted to be reunited with a 'lost' reference in the filing cabinet.

A filing system

As I have just admitted, I do not remember everything and should have some sort of electronic catalogue, but these things take time. My late friend, Ron Pickerill, was far from a computer wizard, but he did have a grant that he used to pay a student to enter new papers into his electronic bibliography of trace fossil publications. If you can find someone else to willingly undertake this task for you – one that I consider onerous, at best – then you are lucky.

What sort of information will you need in your electronic filing system? Essentially, the same data that I have recorded in every reference to any journal article herein. For example:

Donovan, S.K. (2009) A tax on productivity? *Journal of Scholarly Publishing*, **40**: 201–205.

That is, name(s) of author(s), including initials; year of publication; title; where published; volume number; and page numbers. You can embellish this further by including relevant details such as numbers of figures, tables and plates. There are two details that I emphasize here: the importance of where and when a paper was published. Today, it is common practice for journals to include all this bibliographic detail on the first page of each and every paper. In the pre-photocopier days of the past this was not so and, as your collection grows, you will accumulate some rather anonymous papers, published where and published when? This is where bibliographic detective work must come into play. It may be as simple as finding the paper in the reference list of another, more recent paper. Perhaps the typeface is distinctive – remember Sherlock Holmes:

> But this is my special hobby, and the differences are equally obvious. There is as much difference in my eyes between the leaded bourgeois type of a *Times* article and the slovenly print of an evening halfpenny paper as there could be between your Negro and your Esquimaux. The detection of types is one of the most elementary branches of knowledge to the special expert in crime, although I confess that once when I was very young I confused the *Leeds Mercury* with the *Western Morning News* (Doyle, 1981, p. 37).

But can we distinguish a 1923 *Geological Magazine* from the *Annals & Magazine of Natural History*?!

References

Donovan, S.K. (2009) A tax on productivity? *Journal of Scholarly Publishing*, **40**: 201–205.

Donovan, S.K. (2017) *Writing for Earth Scientists: 52 Lessons in Academic Publishing*. Wiley-Blackwell, Chichester.

Doyle, A.C. (1981) [first published 1902]. *The Hound of the Baskervilles*. Penguin, London.

CHAPTER 39

VISITING MUSEUMS

'Well, Challenger, what will you do with your fifty thousand?'

'If you persist in your generous view,' said the Professor, 'I should found a private museum, which has long been one of my dreams' (Doyle, 1912, p. 319).

Not an easy chapter to start. Many of you will look at the title, think, 'You walk in', and wonder how I am going to expand three words to a thousand. Well, 'you walk in' is just the start of the journey.

Displays – the public side

The local history and natural history museums, major and minor, of the British Isles are many and it would be strange if there was not a display of local palaeontology in most, perhaps all of them. I would suggest that the following have some of the major public displays of fossils; if your favourite museum is not on the list, my apologies, but it is meant merely as a guide, certainly not a comprehensive list. In no particular order:

National Museum of Ireland – Natural History Museum

Geological Museum, Trinity College, Dublin

Hunterian Museum, University of Glasgow

National Museum of Scotland, Edinburgh

National Museum Cardiff

Lapworth Museum, Birmingham

Manchester Museum

Sedgwick Museum, Cambridge

Natural History Museum, London

This is a personal list, coloured by my own experiences and visits over the past 40+ years. If your favourite museum fails to make my list, the likelihood is that it lacks a collection of Lower Palaeozoic crinoids to tempt me.

All of these museums have large public displays of fossils of many sorts. These are a boon to the collector wanting to identify their own specimens. After my first day of fossil collecting, ever, in the Gault Clay Formation of Folkestone, Kent, I spent the following Saturday in the late lamented Geological Museum, once a light and informative space to learn (but now a dark and ugly corner of the Natural History Museum).

Having emphasized the displays of a handful of major museums in the British Isles, I note that there are many other museums in many parts of the world with palaeontological exhibitions, many of which have displays of their local geology and palaeontology. It is to these that you should gravitate as a first source to identify your own specimens.

Behind the scenes

No geological museum can display more than a fraction of its holdings. When I worked in the Natural History Museum (NHM) over 20 years ago, there was an agenda by some politicians that every specimen in the collection should be on display – the lot. It takes but little reflection to recognize this as an idea with 'dumb' going through it like 'Blackpool' through a stick of rock. I can understand the sentiment, considering that we were being compared with major art galleries where all of their hundreds of paintings were on display, but, at the time, it was reckoned that the NHM had about 67 million specimens. This included the microscopic – would there be any great public interest in seeing every spore, pollen grain and microfossil on display? These are important for research, but not a major attraction for the public. It was a silly idea spawned from a position of ignorance – that is, a misguided political agenda.

So, there are undoubtedly more specimens in store at your local museum than are on display. How do you get access to them? The first move is to make contact with the curator(s) and explain your interest in the collection. Be specific – for example, you are collecting and studying the brachiopods of the local Jurassic succession; you have referred to the available literature on these species and actually like to compare your identifications with those of the specimens in the museum. All this is reasonable and should elicit a positive response. You will make an appointment, perhaps 'chew the fat' with the curator over a cup of coffee (museum curation and research run on tea and coffee) and then be introduced to the relevant part of the collection. Be armed – have copies of relevant references with you, a good hand lens (the museum may provide a binocular microscope), notebook (either paper or computer or both), pens and pencils. You may be supervised, at least intermittently; once you become a frequent visitor, you may be left to get on with it on your own. The curator is not there to show you how it is done, so be prepared to pursue your own programme of observations.

You are now known to the curator and you have been initiated into the mysteries of the collection. What happens next is largely your decision.

One of the important attributes of any museum specimen is its label(s), which will carry more or less information about the specimen – name, registration number, collector, when accessioned into the collection, locality and horizon. As you are specializing in these species of this age, you may have superior knowledge regarding any of these facets. For example, although a species was originally included in *Agenus*, it may have been reclassified in *Zgenus*. If the classification has not been revised on the label, then either make a neat note of this change on the reverse of the label or include the details on a label-sized slip of paper. Sign and date these additions so that, in the future, others will know the source of these comments.

For example, consider Figure 27.1, an old label with minimal information, but which formed a solid basis for further research. It is a label for a British Geological Survey Geological Survey Museum (prefix BGS GSM) specimen 104611 and 104612 (part and counterpart natural moulds). The collector was Charles W. Peach (1800–1886), who moved from Cornwall to Scotland in 1849, giving a latest date for collection of this specimen (Donovan & Fearnhead, 2017, p. 218). The succession of the Looe Basin is entirely Lower Devonian (Leveridge, 2011, fig. 3), which agrees with the label. *Platycrinus* is a junior synonym of *Platycrinites*. This specimen is certainly *Platycrinites*-like, but it has been transferred to another genus after over 160 years; it is the holotype of *Oehlerticrinus peachi* Donovan & Fearnhead.

You may see fit to donate specimens to one or more museums. Certainly, if you are going to publish and illustrate your specimens, such as in a description of a new species, they will need to be deposited in a recognized museum collection (Chapter 43). Or it may just be that you want a comprehensive set of local specimens with accurate stratigraphic and locality data, etc., to be available for anyone who wants to research using the museum's collection.

No museum ever has sufficient curatorial staff, so perhaps you might like to volunteer your expertise to the museum? Museum volunteers are a valued resource. If you are interested in being involved on a regular basis, then offer your services. I would be surprised if you were not snapped up.

References

Donovan, S.K. & Fearnhead, F.E. (2017) A Lower Devonian hexacrinitid crinoid (Camerata, Monobathrida) from south-west England. *PalZ*, **91**: 217–222.

Doyle, A.C. (1912) *The Lost World: Being an account of the recent amazing adventures of Professor George E. Challenger, Lord John Roxton, Professor Summerlee, and Mr. E.D. Malone of the 'Daily Gazette.'* Hodder and Stoughton, London.

Leveridge, B.E. (2011) The Looe, South Devon and Tavy basins: the Devonian rifted passive margin succession. *Proceedings of the Geologists' Association*, **122**: 616–717.

CHAPTER 40

IDEAS FOR FURTHER INVOLVEMENT

The last few chapters have taken us away from the collection of fossils and into the realm of activities that are inspired by them – attending conferences, reading related magazines, visiting natural history museums and so on. In this chapter I shall look at some facets of palaeontology where you can make a positive impact by bringing fossils to the attention of the uninitiated. We know of the fascination of palaeontology, but there is satisfaction to be had by taking the uninitiated beyond the last *Jurassic Park* movie.

Talking about fossils

Here, I am not considering the research lecture at a palaeontological conference (Chapter 36), but talks on aspects of palaeontology that may be unknown, yet of interest to the general public. That is, talks that introduce the local populace to diverse aspects of the fossil record are popular, particularly if you present some overlooked aspect of the local environment. For example, I have been giving an annual two-part presentation to primary school children at one of the international schools at Amsterdam Zuid. Children of this age are a splendid audience, brimming over with enthusiasm. My first slides remind them of what most people understand when they think of fossils – dinosaurs, extinct Ice Age mammals and prehistoric man. I apologize that none of these groups appear in the rest of this talk and immediately switch to my main focus, the building stones of Amsterdam and other Dutch cities, and the invertebrate fossils in them. This would be an eye-opener to an adult audience. These eight-year-olds lap it up, storing new information with which to stun their families. An important part of the talk is to figure both the fossil invertebrates exposed in two dimensions in building stones and a reconstruction of what they looked like in life (three dimensions).

Part two of my presentation is actually a field class. A week or two after the talk we meet up at Amsterdam Zuid railway station, just a short walk from the school. Here there are raised flower beds with surrounds of polished Mississippian (Lower Carboniferous) limestone slabs at a perfect height for examination by eight-year-olds (Donovan, 2015, fig. 1). They all bring a lens (if they have one), the camera on their smartphone, and a bottle of tap water to splash on the slabs and improve the contrast. The fauna is mixed – solitary and colonial rugose corals, colonial tabulate corals, brachiopods, rostroconchs and crinoid columnals. After an hour of identifying everything, my audience has to walk back to school, with many waves of goodbye, and I have little voice left!

So, that is what I do in a place where there are no natural exposures of rock at the surface. People of all ages, not just infant school children, would appreciate such a talk. I leave you to decide on your subject and your style. Spread the word that you are available to talk and say yes to every reasonable invitation.

Exhibitions

> The exhibits were in the main part, composed of fossil remains from the Cretaceous, Tertiary and Pleistocene Formations, the three notable exceptions being an excellent display of minerals from all over the Commonwealth, by Mrs J. Milbourne, an equally excellent display of Recent Foreign Marine shells, by Miss M. Wheeler and Mr J. Collins, jnr. [– JSHC] and a beautifully mounted collection of British moths and beetles, presented by Mr D. Tulett … A large series of photographs, illustrating the activities of the FGS, taken and mounted by Mr & Mrs R. Everton, occupied most of the available wall space (J.S.H. Collins *in* Donovan & Collins, 2016, p. 95).

This is a report of the first (members only) exhibition of the Freelance Geological Association (28th November, 1953). This led to the first public exhibition a year later (Donovan & Collins, 2016, p. 95, fig. 5A, B). My feeling is that exhibitions by local geological societies may once have been both common and popular, but they have rather faded away – perhaps due to waning public interest in the age of the web? Yet there is at least one major annual exhibition that continues, The Geologists' Association's Festival of Geology, on the first Saturday of November. This is an annual and national event which you can attend and which might give you some useful ideas. Might something on a smaller scale be of interest to the public in your area?

Conference

No, not a rerun of Chapter 36, but, instead, an idea – why not run a conference yourself, under the auspices of your local scientific society? What would be of broad and local interest? Once again there are pockets to pick for ideas. The Manchester Geological Association (MGA) have an annual afternoon meeting, 'The Broadhurst Lectures', named after the late Fred Broadhurst (1928–2009). The MGA decides on a suitable topic for a Saturday afternoon meeting and invites three or four speakers. So, unlike the open conferences in Chapter 36, such a local meeting is more guided to ensure relevant structure and content.

References

Donovan, S.K. (2015) Urban geology: palaeontology at the Wagamama restaurant, Amsterdam Zuid, The Netherlands. *Deposits*, **43**: 8–9.

Donovan, S.K. & Collins, J.S.H. (2016) A brief history of the Freelance Geological Association (FGA), 1948–1967. *Proceedings of the Geologists' Association*, **127**: 90–100.

CHAPTER 41

PUBLISHING I: PERSUADING YOU TO GET INVOLVED

I suggest that your fossil collection might lead to two types of writing, one for your own files and amusement, and the other being published. Both, of course, are ways of writing for yourself, but they vary between having an audience of one (you) and many. Writing can satisfy your desire to record how it happened – how specimens X and Y were collected, and what their significance is to you. If your field notebook is adequate for this purpose, for you, then stick with it. But get your specimens home, perhaps clean them under a tap with an old toothbrush (sometimes the simplest methods are the best), look at them carefully with a lens or microscope and new, perhaps surprising details may become apparent. At the very least, record these details in your register.

Yet it may be that you want to say more and write a separate, extended series of notes as an aid to your memory in the future. Your own records, as such, need only record what you want them to, but they are likely to prove their worth at some future date.

If a specimen interests you, surely it will be of relevance to other collectors, particularly in your local area? There is no reason why you should not spread the word about your specimen by exhibiting it at your local rock and fossil club, showing it to like-minded colleagues and so on. Further, if there is a local geological or natural history journal, it would be common sense to submit an illustrated description of the specimen to them for publication, making your observations concerning its interest and importance available to a wider audience.

I have a lot of time for journals published by local philosophical and scientific societies. When I say local, I recognize that most contributions are geographically constrained. Consider the titles of some of the journals in which I have published in recent years – *The North West Geologist*, *Northumbrian Naturalist*, *Bulletin of the Geological Society of Norfolk* and *Proceedings of the Isle of Wight Natural History and Archaeological Society*, among others. There is no mystery in determining the regional focus of these journals. Many are published annually, some more frequently, some less so.

And just what do I mean by 'local'? None of them are local to me – until recently, I lived in the Netherlands. Yet they are all local to one of my field areas. So, I have two meanings for local. A journal may be relevant to your local area, where you live, or local to your (perhaps more distant) field area.

I emphasize the significance of local journals because, with rare exceptions, they are not commonly published in by professional palaeontologists, who may read them, but less regularly write for them. There are good reasons for this. Academics in universities and museums are expected to publish in high-profile, peer-reviewed journals. Publication in these high-profile journals reflects well on the author's home institution and will be listed in the annual report to the funding body, such as the Government. Failure to publish, and publish regularly, in such journals will likely hinder promotion and may even have an adverse effect on contract renewal; in short, this is the origin of the phrase 'publish or perish'.

I was a full-time academic; I certainly published in peer-reviewed journals, but I was also productive enough to spread my output. I supported, and still support the local journals in my geographic research areas for various reasons. Unlike the big journals, which receive more papers than they can publish, it is common for local journals to aim to cut even, at best. Recently, talking with an editor of a local journal, I discovered that only two papers were in hand for the next issue. I might expect many editors to tell similar tales. Local geological journals need the support of submitted articles from their local collectors, who are logically those who benefit most from the publication of its publication and have most to lose if it is light on copy.

Further, when I publish in a local journal, I am contributing to a resource with the potential to be widely appreciated. If you are going to collect in some new area, then what better home of up-to-date information is there than the local geological journal? A web search may give any number of hits to relevant research papers; but also be sure to curl up with the last few issues of the local journal. For example, when I started to make research trips (and family holidays) to the coast of north Norfolk, I joined the Geological Society of Norfolk and, in doing so, subscribed to their excellent *Bulletin*. Holidays in Norfolk are a thing of the past, but I keep returning to Norfolk for fieldwork (Chapter 47). The *Bulletin* keeps me informed about the latest observations and ideas on the geology and palaeontology of the region.

Taking into account that the target audiences of this book are amateurs and undergraduates, why am I persuading you to write for publication? Surely that is the responsibility of the professional palaeontologists? The simple answer is that I consider all palaeontologists to be capable of making original observations and discoveries. If they do not publish these data, then nobody else will know of them and they will be lost. There are many more things to be discovered about the history of life than there are professional palaeontologists to discover and report them. A committed, non-professional palaeontologist can make a unique and important contribution; many have in the past, and continue to do so. To name just one whom you have met before in these pages, my late friend and co-author, Joe Collins (1927–2019) never had a day's education in a university, but he was nonetheless the leading expert on fossil crabs (decapod crustaceans) and barnacles in the British Isles (Donovan & Mellish, 2020). He was a non-professional palaeontologist with an international profile.

Not everyone can be a Joe Collins, but his dedication is instructive. Like Joe, what you need is a speciality, a corner of creation on which you focus your efforts. This may be geographic (the Lake District, the Norfolk coast), stratigraphic (Upper Ordovician, London Clay Formation) or systematic (Cenozoic gastropods, Palaeozoic corals). The more that you know about your focus, the more you will recognize the gaps in our knowledge. This can only be good for you – what would there be to do if all the questions had already been answered?

In this chapter I have encouraged you to write without discussing the nuts and bolts of authorship. This is too big a subject for a short chapter, but there are very many books on different aspects of the art of academic writing. At the risk of being accused of banging my own drum too hard, Donovan (2017) covers everything that I have left unsaid in this chapter, but takes over 200 pages to do so. If it helps you make the transition from keen collector to published author, I, for one, will be truly delighted.

References

Donovan, S.K. (2017) *Writing for Earth Scientists: 52 Lessons in Academic Publishing.* Wiley-Blackwell, Chichester.

Donovan, S.K. & Mellish, C.J.T. (2020) Mr Joseph Stephen Henry (Joe) Collins, 1927–2019. *Bulletin of the Mizunami Fossil Museum,* **46**: 103–114.

CHAPTER 42

PUBLISHING II: THE HARD WORK OF SELF-EDITING

It is not my intention to be repetitious, so the following is largely rewritten from Donovan (2011, 2017, pp. 81–84). It retains the original structure, but the text is very different while retaining the same messages. The following are my ten rules for academic writing. I intend to avoid too much regurgitation in restating them for the present audience. My rules are a guide that I follow whenever I write, although they are so ingrained that for me they are better called a habit. I trust they are also useful for you. These rules are concerned more with the nuts and bolts of writing rather than the content. Here are some suggestions of how you might enrich your own writing about palaeontology and the writing process.

1 Always carry a notebook

Do not rely on your memory: you will forget. I know this. I have forgotten. Good ideas are ephemeral, so write them down as soon as you have them. I have emphasized the importance of the notebook in the field (Chapter 4), but a different sort of notebook is needed when writing. Even if only grocery shopping, I carry a backpack and in it are a small pencil case (I prefer to write in pencil rather than pen, a personal whim) and a cheap, soft-covered exercise book, about A5 size. This notebook is an essential adjunct to my brain as a writer. I have several of them which, over the years, have filled with scribbled thoughts that I had no intention of misplacing and which later appeared in my publications.

If I infect you with the palaeontological writing bug, you will be delighted to see your articles published, but the most important article is always the next one. That is what you are thinking about. If you have your ideas percolating in the back of your brain, even when you are doing something as mundane as grocery shopping or taking a shower, the right form of words might pop into your head at the least expected moment. Do not lose it, which you will if you fail to write it down immediately.

2 Turn up for work

It is now 6.45 a.m. and I am lurking at home due to the coronavirus. I commonly get up at 5.00 a.m. and catch a train about 6.30 to get to the office about 7.00 a.m. Because I do not have to catch a train at present, I am staying in bed until 5.30 a.m. and, for now, my dining table is my office. Every morning since I became housebound, I have got up every working day and written, mainly this book, but also a trio of research papers that needed to be finished. Yes, an important part of my job as a museum researcher is writing, but just because of an international emergency I am still turning up for work. It has happened before:

> I must also acknowledge the contribution of Hurricane Gilbert [12th September, 1988], which struck Jamaica while I was editing the first chapter and caused suffi-cient destruction that the next six chapters were edited by candlelight (Donovan, 1989a, p. xiii).

After Hurricane Gilbert, at home and in the office, it was five and a half weeks before the electricity was returned. A committed writer (and editor) can work in the worst of situations. We all prefer to work under ideal conditions, but sometimes, like the war correspondent, we must write in less than ideal circumstances. Have a favourite place to write, by all means, but if it is not available, that is not an excuse not to write. Have a timetable for writing and stick to it. Know when your planned writing times occur and use them to write.

3 Protect the time and space in which you write

My scheduled writing time is now and, you will notice, I am using it accordingly. My emails are not open and I am only using the web as a research tool: for example, I used it to check the date when Hurricane Gilbert struck Jamaica (I thought it was the 13th – oops). I am not expecting any phone calls until this evening. As an ex-telephone engineer, I am able to ignore the phone even if it does ring; only very rarely is a call 'urgent'. No music or TV to distract me, ever. Close the door, even lock it. I read somewhere of an author locking herself in the toilet to get some quiet time for writing. You will not write if you are distracted. Do whatever it takes to write in your available conditions of peace.

At this point I make a contradictory admission. I like to write in cafés. If I am surrounded by people who are talking (face-to-face and on their smartphone), eating, playing with the baby and not in any way facilitating my writing, I find it easy to switch them all out. Much of this book was written in the Vascobelo Café near Dam Square in Amsterdam and Subway at Leiden Centraal railway station. These are two places where 200+ words just trip off the pencil over a good breakfast or lunch. I am now comfortable and writing well in my living room, but that just emphasizes that, for me (and you), there are as many potential writing places as you want and need. In all of these places, I write better than in my office at work. It is a fact that I recognize and act on accordingly.

4 Read lots

The theme of this paragraph might be to read everything you can about your favourite group(s) of fossils, but that is only part of the story. In fact, I recommend that you read anything and everything that interests you – novels, biographies, history, whatever. It is all adding to your abilities within the language. I consistently read about ten books per month. Of the 30+ that I have read so far this year, no more than three have been palaeontological or geological. The others have been read for enjoyment and to relish the use of language. The more different ways that you see authors write, the more experience you will have of the possibilities of how to write. Reading is the ideal way of finding out how to write. Whatever you read, you will develop as a palaeontological writer if you are informed and influenced by styles and ideas of other authors.

5 Proceed with care

Earlier this week, I sat down and wrote the first two paragraphs of an introduction to a research paper. The ideas had been buzzing round in my head and I thought I had them organized enough to write a first draft, about 250 words or so. Two or three days later, I typed them up. This move from a handwritten first draft to a typed second draft, following a short break when I did not think about the introduction any more, is part of the evolution of this paper. I saw the text in a slightly different light and made small, but important, changes and improvements. I then printed off what I'd typed and, the next day, read through it with a red pen in my critical right hand. By this stage I am fairly happy with this draft and go on to writing another part of the paper.

Even when I am writing well, I know that I must be cautious. Surplus enthusiasm makes me write too fast and I miss steps in an argument. Taking just a little more time to write anything gives me more chance to recognize frailties in my text. Nobody should be more critical of what I write than I am. A slow, careful approach avoids unnecessary logical slips. Multiple drafts should be considered the norm.

6 The way to write is to actually write

I am going to refer back to 1988, again, and the aftermath of Hurricane Gilbert. I have a fond memory of academics at the Senior Common Room bar at the University of the West Indies in Kingston (UWI), bemoaning the lack of electricity. More than one lamented that they had a paper in their computer that they could finish if only they could turn it on. I kept silent; I still used a manual typewriter and drew my diagrams with ink on tracing paper. With no power for the darkroom, photographs were not possible, so I cut my cloth accordingly and wrote a paper with line drawings only (Donovan, 1989b).

If you want to write then you will write. If you want to talk about writing, as if talking is almost the same as writing, then that is a poor alternative. Similarly, at UWI were the part-time graduate students who will never finish, but who delighted in saying, parrot-like, 'I am writing my Master's thesis' for as long as the university allowed them to pay their fees. Some of them may never have finished a chapter. My advice to potential palaeontological authors – you will get more done if you write rather than talk about it. End of story.

7 Keep a diary

It is here on the desk next to me. This week, apart from anything else, I have been focused on this book. I have written chapters 40 (Monday, when my printer and computer would not speak to each other), 43 (Tuesday), 46 (Wednesday), 45 (Thursday) and 42 (Friday, today, in progress). I should mention that 42, with a coincidental echo from Douglas Adams, is the last chapter of the 52 in *Hands-On Palaeontology*. Next week I shall be writing the 'Introduction', making a PDF of the text and submitting it to the publisher, all guided by my diary.

I use my desk diary to plan, predict (what to do for the next few days is kept up to date) and record my writing days. Organization is the key. My diary keeps me honest and ensures that I remember to do all my necessary tasks, particularly writing, enabling me to balance them over a typical busy and varied working week.

8 If you get stuck, get away from the desk

I did it only this morning. Subheadings 1–5 were written, so I was already halfway through the chapter, but I was yawning, not focusing very well. I slept badly last night; I was awake at 5.10 a.m., still 20 minutes before the alarm clock, not a good sign. So, at 10.30 a.m., I set the alarm clock for 12.30 p.m. and went back to bed for two hours. This afternoon I am revitalized.

Or maybe it is a data processing problem. Your brain needs to be given some air and step back from the writing problem, which it needs to dodge or dance around until a way in can be determined. My ideal space for dealing with a sticky piece of writing is on a walk through the sand dunes on the Dutch coast. A 12 km walk away from most other members of the human race has an effect like putting my brain through a brisk wash cycle. Definitely do not forget the notebook – I've sketched out the guts of an entire article on such a head-clearing walk before now.

9 Editing is everything

I have already introduced this under (4), above. When writing, wear two hats. Writing itself is production, putting the words on paper in the right order, more or less. Editing is quality control, adjusting the words on paper until the order is precise and says exactly what you want them to say. This is important – we often do not write exactly what we think we have written. As your skills as a self-editor improve, you will find it easier to spot where you have been imprecise and the correct form of words will flow more easily.

At this point, recognize that drafts are the key. The first draft will, by definition, be the most shambolic. Your red pen should dance across the page as it corrects facts, format, style, spelling and everything else. Do not despair. This is where you move from the hard donkey work (writing) to your most creative (self-editing). Writing without editing is rare for any article; every paper can be, should be, must be massaged better.

10 Finish what you're writing

I hate an unfinished paper. It signifies nothing. It neither helps nor informs anybody. I live by the words of General Pitt Rivers (1827–1900), 'A discovery dates only from the time of the record of it, and not from the time of its being found in the soil' (quoted in Wheeler, 2004, p. 182).

We all have projects that fail to progress to the finish. I started to write a semi-popular text on Caribbean geology at about the turn of the twenty-first century. It progressed well and then stopped. I did not write a book proposal which, in turn, I did not submit to a publisher. All I had read, written and accumulated for this project sat in a well-labelled box file until about 2016, when I finally asked myself what was the next step? Realism can be brutal – all the paper and the 3½ inch discs were recycled, and I put the box file to a more mundane, yet genuine use.

There will also be joint papers that grind to a halt. There are co-authors that castigate the name of Donovan, S.K., who has sat on a joint paper for years without progress (I have started to photograph some fossil echinoids that I collected in the Dominican Republic in 1997 and will finish the paper this year). There are at least two other recent projects where my part is completed and are now being dealt with by my co-authors. My utopian ideal is no old papers to write up.

Yet new projects are always more exciting than old. The best time to finish any paper is as soon as possible.

References

Donovan, S.K. (ed.) (1989a) *Mass Extinctions: Processes and Evidence*. Belhaven Press, London.

Donovan, S.K. (1989b) The significance of the British Ordovician crinoid fauna. *Modern Geology*, **13**: 243–255.

Donovan, S.K. (2011) Ten rules of academic writing. *Journal of Scholarly Publishing*, **42**: 262–267.

Donovan, S.K. (2017) *Writing for Earth Scientists: 52 Lessons in Academic Publishing*. Wiley-Blackwell, Chichester.

Wheeler, Sir M. (2004) [first published 1954] *Archaeology from the Earth*. Munshiram Manoharlal Publishers, New Delhi.

CHAPTER 43

PUBLISHING III: HOW TO PUBLISH A NEW SPECIES

I believe that I am well qualified to write this chapter. I have described 200+ new species of, mainly, crinoids and trace fossils, both on my own and in association with several co-authors. What can I teach you? You have focused your efforts on a particular group of fossils, collecting them and broadening your expertise by immersing yourself in the literature of that group (Chapters 12, 37, 38). You are an expert by your own efforts. Perhaps you have already 'tested the waters' of publishing, contributing articles to *Deposits* and the journal of your local society. Writing science is not easy, but practice is essential. Many scientists just write research papers, but I have a firm belief that the more different types of article that you write, including book reviews, conference abstracts and so on, the more your own expertise as a writer will blossom (Donovan, 2017, pp. 7–10).

Is your new species based on a specimen(s) that you found in a drawer of your local museum (Chapter 39) or is it defined by one or more that you have collected yourself? If it is already in a museum, then you can refer to its registration number, etc. If not, you will need to donate your type specimen(s) to a museum and to quote their registration number(s) in the paper. This and my other suggestions are in line with the international constraints in zoology (International Commission of Zoological Nomenclature, 1999; for the equivalent botanical code, see McNeill *et al.*, 2012). I also recommend reading Davies (1972, chapter 13) and Winston (1999).

It was not always so. My late friend Stanley Westhead (1910–1986) had more than one type specimen of Carboniferous crinoid in his private collection, the species described by other authors. The Westhead collection is now in the Natural History Museum, London (Donovan, 2012), but not every type specimen in a private collection made a similarly happy transition.

Structure

The assembly of your description of a new species will have a more or less similar structure whether it is a dinosaur, a brachiopod or a fern. The layout will be determined, in part, by the style of the journal, but will be similar to the following.

Genus *Agenus* author, year of publication

Type species: ^^^^

Other species: ^^^^

Diagnosis: ^^^^

Remarks: ^^^^

Range: ^^^^

I am assuming that your new species is a member of a genus that has already been defined; if you are erecting a new genus too, some of the comments regarding a new species (below), such as concerning etymology, can be adapted accordingly. Whatever genus the new species is assigned to, you will have to know its defining characters well. The author and year of publication of *Agenus* will be stated here, and the reference added to the reference list. Every genus is defined by a type species; name it, giving precise bibliographic details, supported by details of its stratigraphic and geographic occurrence. Are there other species in *Agenus*? If there are, say, 20 other nominal species, just say so and refer to the bibliographic reference where they are listed; do not forget to append your new species. If there are only three or four, list them by name with authorship, year of publication and a brief comment on occurrence, such as 'Cretaceous, Maastrichtian, the Netherlands.' The diagnosis is the defining statement regarding *Agenus*, but where do you find it? As a first stop, if it is an invertebrate, go to the relevant part of the *Treatise on Invertebrate Paleontology*, but this may be out of date. Diagnoses evolve as we obtain more complete information regarding a genus based on new and better specimens and investigative techniques. So, search the recent literature for the most current version. Be sure to define the source of your diagnosis, such as '(After Sykes & Milligan, 2017, p. 340.)' Under range the complete geographic and stratigraphic occurrence of the genus needs to be listed and, perhaps, briefly explained.

Agenus aspecies **sp. nov.**

Etymology: ^^^^

Type specimens: ^^^^

Other specimens: ^^^^

Locality and horizon: ^^^^

Diagnosis: ^^^^

Description: ^^^^

Remarks: ^^^^

You will need to give your species a Latinized name: a good dictionary should be an aid in selecting a name and discovering its etymology. You may wish to honour someone, in which case add the suffix –*i*, such as in *westheadi* (male) or –*ae,* as in *fionae* (female).

If you wish to honour the collecting site, add the suffix *-ensis*, such as in *girvanensis*. Whatever your chosen name for your species, called the trivial name, you will need to explain its origins under 'Etymology'.

I have already broached the subject of types (Chapter 7), but it is so important that I make no apologies for treading the same ground. Every species is defined by reference to its type(s). If it is described from a unique specimen, this is the holotype by default, the reference for that species. If you have two or more types, then define the 'most typical' as the holotype and all other members of the type series are paratypes. There may be other specimens that you do not include in the type series for some reason (see, for example, Donovan & Keighley, 2019, under 'Material'). Locality and horizon are concepts that will be well known to you by now. The genus *Agenus* was defined above; the new species *Agenus aspecies* needs to be diagnosed within *Agenus*, defining those features that separate it from all other species in this genus. This is succinct; in contrast, the description needs to be comprehensive, defining all the features of *Agenus aspecies*, whether they are unique or shared by a hundred other species. The 'Remarks' section gives you elbow room to establish your new species. Explain why it is a member of *Agenus* and not, say, *Bgenus*. How does it differ from the other species of *Agenus*? What other observations of significance can be gleaned from the specimen that were not emphasized in the 'Description', such as its preservation? For a recent example of such a discussion, see Fearnhead *et al.* (2020).

References

Davies, A.M. revised Stubblefield, J. (1972) [tenth impression] *An Introduction to Palaeontology*. Third edition [edition first published 1961]. Thomas Murby, London.

Donovan, S.K. (2012) Stanley Westhead and the Lower Carboniferous (Mississippian) crinoids of the Clitheroe area, Lancashire. *Proceedings of the Yorkshire Geological Society*, **59**: 15–20.

Donovan, S.K. (2017) *Writing for Earth Scientists: 52 Lessons in Academic Publishing*. Wiley-Blackwell, Chichester.

Donovan, S.K. & Keighley, D.G. (2019) A 'British' Silurian crinoid from Quinn Point, New Brunswick, eastern Canada: Designation of types. *Proceedings of the Geologists' Association*, **130**: 770–771.

Fearnhead, F.E., Donovan, S.K., Botting, J.P. & Muir, L.A. (2020). A Lower Silurian (Llandovery) diplobathrid crinoid (Camerata) from mid-Wales. *Geological Magazine*, **157**: 1176-1180.

International Commission on Zoological Nomenclature. (1999) *International Code of Zoological Nomenclature*. Fourth edition. International Trust for Zoological Nomenclature, London.

McNeill, J., Barrie, F.R., Buck, W.R., Demoulin, V., Greuter, W., Hawksworth, D.L., Herendeen, P.S., Knapp, S., Marhold, K., Prado, J., Prud'homme Van Reine, W.F., Smith, G.F., Wiersema, J.H. & Turland, N.J. (2012) *International Code of Nomenclature for Algae, Fungi, and Plants (Melbourne Code) adopted by the Eighteenth International Botanical Congress Melbourne, Australia, July 2011*. Regnum Vegetabile 154, A.R.G. Gantner Verlag KG.

Winston, J. (1999) *Describing Species: Practical Taxonomic Procedure for Biologists*. Columbia University Press, New York

FOSSILS IN MANY FIELDS

CHAPTER 44

THE FIELD GUIDE

My apologies, but I regard the field guide as possibly the most important form of geological/palaeontological publication for the amateur and, in consequence, I intend to ride it until the end of this book. I maintain that, apart from the geological map, the field guide is the most geological of publications. It follows from this that, in the absence of something that we might call a palaeontological map, a palaeontological field guide is the most palaeontological of publications. I am therefore finishing this book with a eulogy to field guides for the palaeontologist, and follow it with a series of guides to places that I know reasonably well. The latter serve at least three purposes: they will introduce you to the use of a field guide to find fossils; I intend to demonstrate how you can find fossils almost anywhere if you know what to look for; and I hope that they will encourage you to write your own field guides.

Field guides are of use to the novice and the experienced palaeontologist alike. They have a different purpose to the research paper. I can best illustrate this by reference to some of my own publications. The most impressive fossiliferous site that I ever worked on is a bed of the giant oyster *Crassostrea virginica* (Gmelin) in the Neogene on the central

Figure 44.1 The author (*c.* 1987) at the *Crassostrea virginica* bed, Farquhar's Beach, parish of Clarendon, south-central Jamaica. The beds dip to the right; the base of the *C. virginica* bed is about the level of the author's knees and is *c.* 3.3 m in thickness. See also Figure 19.1.

Figure 44.2 Detail of the *C. virginica* bed showing *in situ* oysters up to 160 mm in length.

south coast of Jamaica (Littlewood & Donovan, 1988; Figs 19.1, 44.1, 44.2 herein). The original description of this bed included details such as measured section and photographs, as well as a locality map. As comprehensive as this is, our research paper focused on just one part of a coastal section that is *c*.2 km long, and includes an array of sedimentary, structural and palaeontological features. To examine the whole section, you would likely refer to Donovan *et al.* (1995, pp. 7–12), which grew out of a report of a field meeting (Donovan *et al.*, 1989). Later, in 1998, I was again co-leader of a field meeting of the Geological Society of Jamaica to this section, which led to a published report (Donovan & Miller, 1999). This last publication is the most comprehensive, although there are still aspects of this section awaiting further investigation. The measured sections and itinerary of Donovan & Miller (1999) would be of use, but, sadly, the reproduction of the photographs is poor. (The originals were good; we even hired a fishing boat to photograph the section from offshore.)

As you become more acquainted with your favourite area for collecting fossils, you will be developing the necessary information for your own field guide, both in your head and in your field notebook. It will take a little time, but the more time you spend in the field, the more observations you will make. This is the bed that always yields at least a few fine brachiopods; 40 m north-west along the beach and near to the wreck is a ledge made by a bed that stands out from the cliff face, and is rich in sponges and high-spired gastropods; and then up the coast path to the edge of the fields where fine ammonites are exposed by the plough – they are rare, but are worth finding. You will develop an expertise that nobody else possesses. This is one of the most invigorating aspects of field palaeontology – developing your knowledge base of your field area.

Even if others are collecting in the same beds, they do not observe in quite the same way that you do and will not make exactly the same collection of fossils as you. Chatting with fellow collectors in the field will cross-fertilize their knowledge with yours – in effect, they are likely to provide information that you lacked, until now.

What should you do with your unique knowledge of your field area? You could just keep it to yourself. This happens more often than you might think. As an analogous example, but from a different situation, a Ph.D. in geology or palaeontology at a British/Irish/North American university requires the candidate to write a thesis, a weighty tome in one or more volumes, but not that it is published. That is, after three years' study, supported by a research council, the university, industry, or mum'n'dad, there is no requirement to publish. There are untold gems in university libraries as you read this, bursting with information that will never be made widely available by publication.

So, staying quiet is an option, but not one of which I approve. I want you to get your ideas and observations out in the open. For example, why not lead a field excursion for your local geological society? They will likely jump at your offer. The society's field excursion secretary will be in search of new field trips for this or next summer, and you can step into the breach with confidence. You should write a brief outline of the day for the attendees – say, one page of text and a map – and pay careful heed in the field to their observations and discoveries. I guarantee that if you take a dozen keen collectors to any of your sites, be they professionals, amateurs or students, something will be collected or observed that you have not seen hitherto. You may be the leader of the trip, but expect the attendees to educate you while you educate them. The more pairs of eyes on the rocks, the better will be the results.

Afterwards, congratulations from me on a trip well run. Does it stop there? Well, it can, but why not complete the job? You are in a strong position to write up a report of the field meeting for your local geological society's newsletter or, better, in more depth in the journal. It can be based on the brief outline that you wrote for the attendees, supplemented by more details of the geology and a report of what happened on the day ('The sharp eyes of Mrs Smith's son led him to a most unusual specimen …'). A report of your field meeting will be used by interested individuals as a field guide in the future. Alternatively, your local society might publish a field guide in their journal before the meeting. This can be supplemented by a report of the field meeting after the event. These two publications on one site should not be considered overkill, providing you are not overly repetitive: they can provide mutually supportive, yet different, accounts of the same area.

References

Donovan, S.K., Jackson, T.A., Dixon, H.L. & Doyle, E.N. (1995) *Eastern and central Jamaica.* *Geologists' Association Guides*, **53**: i+62 pp.

Donovan, S.K., Jackson, T.A. & Littlewood, D.T.J. (1989) Report of a field meeting to the Round Hill region of southern Clarendon, 9 April 1988. *Journal of the Geological Society of Jamaica*, **25** (for 1988): 44–47.

Donovan, S.K. & Miller, D.J. (1999) Report of a field meeting to south-central Jamaica, 23rd May, 1998. *Journal of the Geological Society of Jamaica*, **33** (for 1998): 31–41.

Littlewood, D.T.J. & Donovan, S.K. (1988) Variation of Recent and fossil *Crassostrea* in Jamaica. *Palaeontology*, **31**: 1013–1028.

CHAPTER 45

FIELD TRIP: DEN HAAG, THE NETHERLANDS

Preamble

Chapters 45 and 46 are my COVID19 contributions. I had intended to get into the field and write two fresh field guides of different areas when I, like most of the rest of the world, was confined to home. So, nothing new was favoured by circumstances and I had to juggle some chapters. The present guide is based on Donovan (2019).

This chapter has two purposes: to demonstrate how fossiliferous rocks and fossils are integral to our modern built environment; and to encourage you to actively search for them. I am primarily considering the urban environment. The fossils that I am discussing are mainly invertebrates, although trace fossils, plants and even disarticulated vertebrate bones can make rare appearances. Many invertebrates are seen in facing and building stones, and in street furniture, in Dutch and British cities, and elsewhere. That these fossils are not more widely recognized by the man/woman in the street is explained by their relative lack of expertise, but geologists are also inclined to ignore them.

The reasons for this indifference or ignorance are surely several. To our man/woman in the street, a fossil is a big bony thing – a mammoth or dinosaur or Neanderthal Man – whereas invertebrates may be considered trivial, perhaps, and are forgotten. Fossils are three-dimensional, something to be walked around in a museum, yet, in truth, most fossils are smaller than a fist. Further, a fossil in a building stone is a two-dimensional slice through a three-dimensional object and may appear more like a rune or a sport of nature than the skeleton of an animal that died 300 million years ago. Taken together, these criteria make the fossil in a building stone almost invisible and partly indistinguishable, even to an informed geologist who chooses not to observe too closely.

So, how might you go about fossil hunting in the streets of your village, town or city? Equipment need only be basic: a hand lens; a notebook and pencil; a soft brush and a bottle of water for cleaning surfaces; a camera; and a street map should suffice. No hammers, please. All of your collecting will be by camera, hardly an extravagance in the age of digital photography. And have your wits about you.

How to get there

The question of where to start is not trivial. It may be complicated or it can be easily solved – just go out of your front door and carry on walking wherever your feet take you. I have been particularly interested in Mississippian fossiliferous limestones in Dutch towns and cities for many years – Amsterdam, Leiden, Hoofddorp, Maastricht and Utrecht have all been the subject of my camera and my pen. So, for this study, I wanted something fresh, not recycled. I chose a city that I know only a little, Den Haag (= The Hague). This proved to be an inspired choice, and a wealth of specimens were found on one long Monday morning before lunch. My starting point was Den Haag Centraal Station (CS), the main railway terminus of the city.

Geological history

It would be impossible to discuss each and every type of fossil that you might encounter in building stones, so I am going to be selective and focus on one stratigraphic interval. I take the widespread Mississippian (Lower Carboniferous) limestones of Den Haag as my focus and will consider some general principles, from which the reader can extrapolate to rocks of different age(s) in their home town. The Mississippian includes some of the world's thickest and most extensive limestone deposits across North America – Europe – Asia, resources which are important both as building stones and for cement manufacture. These limestones have not just been used locally in buildings, but they have been imported. There is no exposure of Mississippian limestones in the Netherlands, yet these rocks are widespread in the built environment.

What to look for

Identifying fossils in building stones is to recognize three-dimensional objects from random two-dimensional sections. Some fossils are easy to spot, others are more ambiguous. The notes and figures below are intended to facilitate the identification process, as will the specialist references (Anon, 1969; Donovan, 2016; Reumer, 2016; Donovan et al., 2017; Donovan & Wyse Jackson, 2018; Donovan & Harper, 2018; Van Ruiten & Donovan, 2018): but expect the unexpected.

Corals (Figs 45.1, 45.2)

Mississippian fossil corals should be among the easier groups to identify. They are solitary or, more commonly, colonial; rugose corals have radiating internal septa, resembling the spokes of a bicycle in two dimensions (Fig. 45.1), but they are actually plates that supported the folded gut in life. Tabulate corals are always colonial and lack septa (Fig. 45.2). Colonies may be massive and domed or branched. Both groups have a calcite skeleton, in contrast to the aragonitic hard parts of extant scleractinian corals.

Brachiopods (Fig. 45.3)

Brachiopods were a common part of the Mississippian marine fauna and are a diverse component of many building stones. They are bivalved and are a more common component of Mississippian rocks than bivalve molluscs. A disarticulated valve in section may be rather nondescript, either rounded or comma-like depending on the angle of cut. Valves that are wider than long and have a straight hingeline are probably spiriferids.

[Left] Figure 45.1 Solitary rugose coral in transverse section showing the radial septa; Departement van Justitie building, Den Haag, Netherlands. Scale in cm.

[Right] Figure 45.2 *Michelinia* sp., a tabulate colonial coral in transverse section; Departement van Justitie building, Den Haag, Netherlands (after Donovan, 2019, fig. 3c). Scale in cm. Also see Figure 10.1.

[Left] Figure 45.3 Inverted life assemblage of productid brachiopods; Departement van Justitie building, Den Haag, Netherlands (also figured in Donovan, 2019, fig. 3h). Although a life assemblage, the block was mounted upside down by the builder. Scale in cm.

[Right] Figure 45.4 Crinoid columnals in an etched limestone; Langer Vijverburg 8, Den Haag, Netherlands (also see Donovan, 2019, fig. 5b). The specimen in the centre is bored, *Oichnus* isp. Scale in cm.

Articulated productid brachiopods are particularly distinctive, having a concavo-convex shell rather than the much more common biconvex arrangement of the valves (Figs 24.2, 45.3). In transverse section one valve is seen inside the other; in longitudinal section, the shell is a double-walled comma-like shape.

Crinoids (Fig. 45.4)

I have spent many hours scanning Mississippian limestones in the Netherlands for articulated crinoid cups and crowns; they are very rare (Donovan, 2020), yet disarticulated fragments of crinoids are abundant. An unexpected specimen is pitted (Fig. 45.4) – these are probably borings and are assigned to the ichnogenus *Oichnus* Bromley.

Molluscs (Figs 45.5, 45.6)

Molluscs come in many shapes and forms. Two groups are apparent in these rocks, namely planispiral shells (gastropods? cephalopods?) (Fig. 45.5) and rostroconchs. The planispiral shells are a puzzle. Lacking internal septa, my assumption is that they are gastropods such as *Straparollus* sp., but some authors consider them coiled cephalopods (ammonoids or nautiloids) in which the septa are not preserved. Why they are not preserved is a question that I cannot answer, hence my preference.

Rostroconchs (Fig. 45.6) are extinct univalves that are reminiscent of bivalves in external appearance. They have only recently been recognized in limestones in the Netherlands (Donovan & Madern, 2017). They are an example of a group that, now my collector's 'search pattern' has been set, I seem to find everywhere and anywhere.

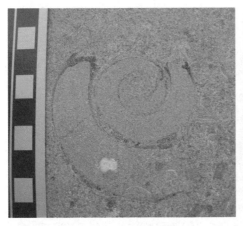 

[Left] Figure 45.5 Planar coiled mollusc, either gastropod or cephalopod; Parkstraat 83–91, Den Haag, Netherlands (after Donovan, 2019, fig. 7b). Scale in cm.

[Right] Figure 45.6 Rostroconch (probably upside down); Parkstraat 83–91, Den Haag, Netherlands (after Donovan, 2019, fig. 7d). Scale in cm.

Bryozoans (Fig. 45.7)

Bryozoans are a common component of the Mississippian fauna, but are generally preserved as small fragments, difficult to recognize and even harder to identify. The figured specimen (Fig. 45.7) is part of a fenestrate bryozoan colony that would have been somewhat bigger and vase-shaped when complete (Donovan & Wyse Jackson, 2018). Note the window-like openings through the colony; hence it is fenestrate from the Latin *fenestra*, meaning window.

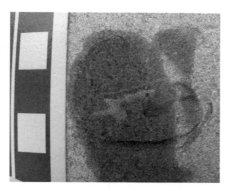

Figure 45.7 Fragment of a fenestrate bryozoan colony; Parkstraat 83–91, Den Haag, Netherlands (after Donovan, 2019, fig. 7f). Scale in cm.

Trace fossils (Fig. 45.4)

Mottled surfaces, presumably cut parallel to bedding, are intriguing. They are seen in both facing stones, where differences in colour are apparent, and on steps, where the tread of many feet has produced an unevenness. This effect might have been the product of *Thalassinoides* Ehrenberg (Donovan, 2019, figs 5c, 7e) and this ichnogenus is known from the Mississippian, but it is commoner in the post-Palaeozoic.

For a discussion of small round holes in a crinoid stem (Fig. 45.4), see above. Again, I have examined many hundreds of crinoid ossicles in building stones and this is the first specimen I have seen bored, *Oichnus*? isp. The dense cluster of many pits is unusual and begs explanation.

References

Anon. (1969) *British Palaeozoic Fossils*. Third edition. Trustees of the British Museum (Natural History), London.

Donovan, S.K. (2016) A mollusc–coral interaction in a paving slab, Leiden, the Netherlands. *Bulletin of the Mizunami Fossil Museum*, **42**: 45–46.

Donovan, S.K. (2019) Urban geology: Mississippian in the Mainstreet. *Geology Today*, **35**: 135–139.

Donovan, S.K. (2020) Early Carboniferous (Mississippian) crinoids from Eindhoven, the Netherlands. *Bulletin of the Mizunami Fossil Museum*, **46**: 5–9.

Donovan, S.K. & Harper, D.A.T. (2018) Urban geology: productid brachiopods in Amsterdam and Utrecht. *Deposits*, **56**: 10–12.

Donovan, S.K., Jagt, J.W.M. & Jagt-Yazykova, E.A. (2017) A well-preserved crinoid stem in a building stone (Lower Carboniferous, Mississippian) at Maastricht, the Netherlands. *Bulletin of the Mizunami Fossil Museum*, **43**: 23–25.

Donovan, S.K. & Madern, P.A. (2017) Rostroconchs in Leiden. *Swiss Journal of Palaeontology*, **135**: 349–352.

Donovan, S.K. & Wyse Jackson, P.N. (2018) Well-preserved fenestrate bryozoans in Mississippian building stones, Utrecht, The Netherlands. *Swiss Journal of Palaeontology*, **137**: 99–102.

Reumer, J. (2016) *Kijk waar je loopt! Over stadspaleontologie.* Historische Uitgeverij, Netherlands.

Ruiten, D.M. van & Donovan, S.K. (2018) Provenance, systematics and palaeoecology of Mississippian (Lower Carboniferous) corals (subclasses Rugosa, Tabulata) preserved in an urban environment, Leiden, the Netherlands. *Bulletin of the Mizunami Fossil Museum*, **44**: 39–50.

CHAPTER 46

FIELD TRIP: THE PILTDOWN TRAIL

Preamble

(Condensed from Donovan, 2015.) Even though it is within easy reach of London, it is hardly commonplace for geologists from the city to visit the type locality of *Eoanthropus dawsoni* Woodward *in* Dawson & Woodward, 1913, or, as it is better known, the discredited Piltdown Man (Fig. 46.1). Yet, when first 'discovered', the site at Barkham Manor near Piltdown, East Sussex, was a focus for a great field trip (85 official attendees, but perhaps over 100 in total) organized by the Geologists' Association on 12th July, 1913, and led by the experts, Charles Dawson and Arthur Smith Woodward (Anon., 1914, p. 277; Walsh, 1996, pp. 39–40, 5th plate (unnumbered); Russell, 2003, p. 222, fig. 98). Yet despite Piltdown Man being exposed as a forgery, this site had been the jewel in the crown of British palaeoanthropology for 40 years (Prosser, 2009). In providing this field guide, it is emphasized that this is an excursion into the history of palaeontology, but there are no fossils to be collected.

Herein, the three Piltdown sites are called Netherhall Farm (Locality 46.1), Barkham Manor (Locality 46.2, the type locality) and Barcombe Mills (Locality 46.3). These correspond to Piltdown Man II, I and III, respectively, of Russell (2003) and others. Piltdown Man III is a misnomer, as there is no suggestion that the skull fragments from this site are anything other than *Homo sapiens*; it is better referred to as Barcombe Mills Man. The sites are visited in an order that

Figure 46.1 'The first Englishman Piltdown, Sussex' (after Baikie, 1928, frontispiece). With a spear in one hand and the 'cricket bat' on a rope in the other, Piltdown Man stalks a beaver, *Castor* sp.

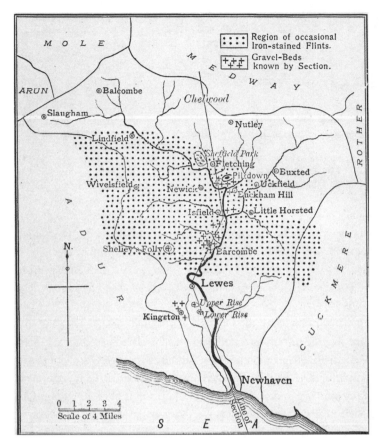

Figure 46.2
'Plan of the basin
of the Sussex
Ouse, showing
the distribution
of iron-stained
flints and flint-
bearing gravels'
(after Dawson
& Woodward,
1913, fig. 1).

will facilitate the walker, who can start at Sheffield Park railway station and finish at a bus stop with easy egress to Uckfield and Lewes. Recommended maps include the Ordnance Survey 1:25,000 Explorer series, numbers 122 *Brighton & Hove*, 123 *Eastbourne & Beachy Head* and 135 *Ashdown Forest*.

Geological history

There are two geologists whose names are associated with the Piltdown Man sites at Barkham Manor and elsewhere, namely the amateur Charles Dawson, FSA, FGS (1864–1916), and Francis Hereward Edmunds, MA, FGS (1893–1960), of the Geological Survey of Great Britain. Dawson, as 'discoverer' of the skull at Barkham Manor and perpetrator of the forgery, is already well served biographically (such as Walsh, 1996; Russell, 2003; Donovan, 2016). His contribution to the geology seems to have been worthy. It included a map showing the distribution of iron-stained flints and flint-bearing gravels in the region of the River Ouse (Fig. 46.2) and an accurate measured section of the gravels at Barkham Manor.

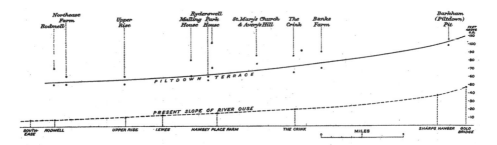

Figure 46.3 'Diagram to show the relationship of the Piltdown
Terrace to the River Ouse' (after Edmunds, 1926).

Francis Edmunds is the 'forgotten man' of the Piltdown controversy. Edmunds took
first class honours in geology, part II, in the Natural Science Tripos at Emmanuel College,
Cambridge, in 1922. He was employed by the Geological Survey of Great Britain from
1922 until 1957. Edmunds was author of *British Regional Geology: The Wealden District*
(Edmunds, 1935, 1948, 1954).

Edmunds contributed to the geology of the Piltdown Man I site (= Locality 46.2
herein) by demonstrating that the terrace level had been miscorrelated based on Dawson's
original estimate of elevation. The determination by Edmunds suggested that the remains
were younger than determined from morphology and prejudice ('British naturalists
accepted the Piltdown remains as genuine – perhaps on the grounds of national prestige';
Bowler, 1986, p. 36). The conclusions of Edmunds were largely ignored until the forgery
was recognized (see Edmunds, 1950, 1955).

The diagram of the relationship of the so-called Piltdown Terrace to the River Ouse
by Edmunds (1926; Fig. 46.3 herein) shows it to be less than 70 feet above the present
river level, that is, less than about 22 m; Edmunds (1950, p. 133) later calculated it was of
50—55 feet (about 15—17 m). This is in contrast to the original determination of Dawson
(in Dawson & Woodward, 1913, p. 119) that 'The gravel … lies about 80 feet [25 m] above
the level of the main stream of the Ouse.' This error was repeated as late as 1948. Dawson's
figure made the Piltdown remains to be older than Edmunds' estimate and was widely
adopted.

What to look for

All of my field guides can be completed by car, but I much prefer rail. Trains are more
comfortable than cars with more leg room (I am 1.92 m tall); you can read and prepare
for the day ahead; and enjoy the scenery, always of interest to the geologist.

Locality 46.1: Netherhall Farm, near Sheffield Park

National rail from London Victoria will take you to East Grinstead. Change onto the

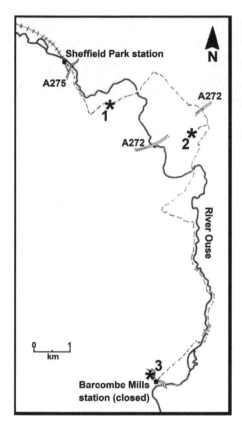

Figure 46.4 Outline map of field excursion described herein and best used in conjunction with Ordnance Survey sheets recommended in the text (after Donovan, 2015, fig. 9). Key: * = localities numbered as in the text; filled boxes = railway stations (including the closed Barcombe Mills station); solid black line = River Ouse; dashed line = recommended route; trellis = railway. Two A-roads are marked where they cross the route. To catch a bus to Uckfield or Lewes railway stations at the end of the walk, follow the road past the closed Barcombe Mills station to the south-east; at the A26 road junction, turn left for bus stops.

Bluebell Railway to Sheffield Park (Fig. 46.4). Leave the station by the approach road and cross the main road (A275). Follow the footpath sign to the Sussex Ouse Valley Way (SOVW) and walk on the floodplain of the Ouse (river to your left). Follow the SOVW and turn up the valley side through Rotherfield Wood. At the top of the path, go through the gate and turn left on the minor road towards Fletching. This road more or less follows the contour; note exposures of Wealden sandstones in cuttings on this road and elsewhere. Netherhall Farm is about halfway from the SOVW gate to Fletching (Fig. 46.4) [NGR TQ 416 226]. (Sheffield Park station to Netherhall Farm about 2.4 km.)

This is the least well-defined of Dawson's hominin sites geographically. It is not certain that Netherhall Farm is the correct site of Piltdown II. That there was a professional link between Dawson the solicitor and Netherhall Farm, analogous to his link with Barkham Manor that led to the discovery of the Piltdown Man type locality (= Locality 46.2), may have been the only evidence available (Walsh, 1996, p. 231, note 58; Russell, 2003, p. 241). Netherhall Farm is easily found and the surrounding fields form an obvious terrace above the River Ouse, but there is no obvious sign of a disused gravel pit.

Figure 46.5 Locality 2, Barkham Manor, Piltdown I site (after Donovan, 2015, fig. 11). **(A)** The view through the gates at Barkham Manor, looking towards the house. The Piltdown I site is over the crest of the hill and to the right of the track. **(B)** Memorial stone with the author (1.92 m) for scale. The inscription states: 'Here in the old river gravel Mr. Charles Dawson F.S.A. found the fossil skull of Piltdown Man 1912–1913. The discovery was described by Mr. Charles Dawson and Sir Arthur Smith Woodward in the Quarterly Journal of the Geological Society 1913–1915.' **(C)** The site of the excavation today; memorial stone to right. Barkham Manor is in the left distance.

Locality 46.2: Barkham Manor, Piltdown

Continue walking north-east towards Fletching (Fig. 46.4). The flat river terrace is particularly apparent shortly after Netherhall Farm [NGR TQ 419 227] to the left of the road. However, as already stated, there is no sign of a disused gravel pit. The road drops down into the valley of the Ouse, which is crossed by Fletching Mill Bridge, and rises up the opposite side. A disused pit here is in Wealden.

At the parish church of St Andrew and St Mary the Virgin, Fletching, turn right into Church Street, signposted prophetically 'Piltdown I' (=1)! The road gently undulates, sloping down to cross a tributary to the Ouse and gradually climbing to the junction with the A272. Cross this main road and enter Lodge Lane, opposite. At the bottom of the lane turn left towards Uckfield. After about 100 m take the right turn by the pond, signposted to Barcombe. After 200 m take the right turn into Barkham Lane. Barkham Manor is on the left (Fig. 46.5), protected by sturdy iron gates and a sign 'Private No Trespassers' [NGR TQ 441 219 at gates]. (Netherhall Farm to Barkham Manor gates about 4.0 km.)

The most significant of Dawson's sites – the type locality of *Eoanthropus dawsoni* – is also the most difficult to visit. To view the site, advance arrangements should be made with the current occupants of the property. It has been so well described hitherto that it seems irrelevant to say more, apart from directing the reader to Prosser's (2009) account.

Locality 46.3: Barcombe Mills

Return to end of Barkham Lane and turn right on the minor road towards Barcombe (Fig. 46.4). Note how extensive the flat plateau/terraces are in this area, dissected by various river valleys. From Moon's Farm the road drops gently downslope towards Sharpsbridge and the Ouse valley; the pillbox in the field on the right, before the bridge, is a distinctive landmark. The climb is steeper on the other side.

The SOVW crosses the road just after Broomlye Cottage; turn left (east) onto the footpath that gently drops down towards the river. On reaching the floodplain of the River Ouse, turn south on the SOVW towards Barcombe Mills; note the ox bow lakes in various stages of infilling and overgrowth. Another feature of this walk is the number of pillboxes lurking in vegetated hiding places. After passing under the disused railway bridge, take the next footbridge onto Anchor Lane (the Anchor Inn on the river is a good stop for a late lunch). Continue down Anchor Lane and turn left onto the closed railway, a bridleway, which will take you directly to the site of the former Barcombe Mills station [NGR TQ 429 149], now private houses. (Barkham Manor gates to former Barcombe Mills station about 10.0 km.)

There is only one hill above the site of the station at Barcombe Mills, to the north-west. The road from the station site to Barcombe turns a dog leg on the hill [NGR TQ 428 151]. This site was not collected just by Dawson, but apparently also by Smith Woodward and Teilhard de Chardin (Russell, 2003, pp. 232 and 233, respectively). Sadly, nobody appears to have left a marked map or a precise description of the locality. However, a letter by Dawson to Smith Woodward suggests that what specimens exist were, allegedly, collected by him in the float from the field.

I have written this guide for the walker, as this is the best way to appreciate the scenic beauty of the valley of the River Ouse and its terraces. It is the Ouse and its terraces that form the main geological story of Piltdown, and this guide is designed to show what is apparent at the present day, as well as the sites that were, for 40 years, amongst the most important in all British palaeoanthropology.

A distinctive feature of Charles Dawson's forgeries was his vagueness – what Russell (2012, p. 40) referred to as the 'standard tactic: obscuring the exact nature of both find and provenance.' Herein, I have reworked the published accounts and ideas to produce the best possible guide to these historically important sites.

References

Anon. (1914) The Annual Report of the Council of the Association for the year 1913. *Proceedings of the Geologists' Association*, **25**: 271–279.

Baikie, J. (1928) *Peeps at Men of the Old Stone Age*. A. & C. Black, London.

Bowler, P.J. (1986) *Theories of Human Evolution: A Century of Debate, 1844–1944*. Johns Hopkins University Press, Baltimore.

Dawson, C. & Woodward, A.S. (1913) On the discovery of a Palaeolithic human skull and mandible in a flint-bearing gravel overlying the Wealden (Hastings Beds) at Piltdown,

Fletching (Sussex). *Quarterly Journal of the Geological Society, London*, **69**: 117–151.

Donovan, S.K. (2015) A field guide to Charles Dawson's discredited sites implicated in the Piltdown hoax. *Proceedings of the Geologists' Association*, **126**: 599–607.

Donovan, S.K. (2016) The triumph of the Dawsonian method. *Proceedings of the Geologists' Association*, **127**: 101–106.

Edmunds, F.H. (1926) Fig. 10. *In*: White, H.J.O. *The Geology of the Country near Lewes. Explanation of Sheet 319*. Memoirs of the Geological Survey, England. H.M.S.O., London.

Edmunds, F.H. (1935) *British Regional Geology. The Wealden District*. HMSO, London.

Edmunds, F.H. (1948) *British Regional Geology. The Wealden District*. Second edition. HMSO, London.

Edmunds, F.H. (1950) Note on the gravel deposit from which the Piltdown skull was obtained. *Proceedings of the Geological Society, London*, **106**: 133–134.

Edmunds, F.H. (1954) *British Regional Geology. The Wealden District*. Third edition. HMSO, London.

Edmunds, F.H. (1955) The geology of the Piltdown neighbourhood. *In*: Weiner, J.E. and eleven others, Further contributions to the solution of the Piltdown problem. *Bulletin of the British Museum (Natural History)*, Geology series, **2**: 273–275.

Prosser, C. (2009) The Piltdown skull site: the rise and fall of Britain's first geological National Nature Reserve and its place in the history of nature conservation. *Proceedings of the Geologists' Association*, **120**: 79–88.

Russell, M. (2003) *Piltdown Man: The Secret Life of Charles Dawson & the World's greatest archaeological Hoax*. Tempus, Stroud.

Russell, M. (2012) *The Piltdown Man Hoax: Case Closed*. The History Press, Stroud.

Walsh, J.E. (1996) *Unravelling Piltdown: The Science Fraud of the Century and its Solution*. Random House, New York.

CHAPTER 47

FIELD TRIP: OVERSTRAND TO CROMER, NORFOLK

Preamble

The walk along the beach in north Norfolk from Overstrand to Cromer and return is one of my favourite beachcombing routes for collecting Chalk fossils, Recent marine borings and reworked erratics. It has yielded more than a few interesting specimens (Donovan & Lewis, 2010, 2011; Donovan, 2010, 2011a, b, 2012, 2013, 2017; Donovan *et al.*, 2019). The route itself is most pleasant, more or less straight (Fig. 47.1), with the North Sea on one side and the cliffs of north Norfolk on the other. If you are surprised by the cliffs – 'Very flat, Norfolk' as Noel Coward said in *Private Lives* – then I can say that this is the hilly end of the county, with the Cromer Ridge produced by glacial action (see below). I tend to walk one way in the morning, have lunch in Cromer (usually local fish) and walk back. Apart from the good exercise, the clasts on the beach look different in the afternoon light and walking in the opposite direction. I almost invariably find specimens returning to Overstrand from Cromer that must have been there in the morning, but which were not obvious from a different angle and with different illumination.

This is also an excellent holiday destination. I started collecting around Overstrand while on family holidays, the children enjoying themselves in the sea while Dad looked at stones on the beach. So, why not take the family, too?

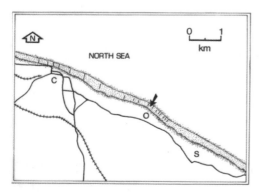

Figure 47.1 Outline map of the north coast of Norfolk between Cromer (C), Overstrand (O) and Sidestrand (S), after Donovan (2010, fig. 1). The dark arrow indicates the author's point of access to the beach. The stippled area is between the low water mark and cliff top; it includes both the beach (groynes are indicated) and slope of the cliffs. Principal roads are shown as solid lines; railways are shown as trellised lines.

How to get there

I recommend taking the stopping train from Norwich to Cromer, which runs about once per hour. London to Norwich is quick and easy, but trains from elsewhere can take an age. I am usually coming from Manchester, and that journey takes about four and a half hours; onto this add Norwich to Cromer, another hour. I prefer to stay in Overstrand, which is a short ride by bus or taxi from Cromer station. Once I am installed, fieldwork is mainly on foot, but it is useful to know something of the local buses and taxis. I always stay at The White Horse in Overstrand, which is always friendly and has many advantages for a base, not least the excellent kippers served at breakfast and the close access to the coast.

Geological history

(After Donovan & Lewis, 2011, pp. 44–45.) 'It is probable that the chalk and flint cobbles described from the beach at Overstrand are local in origin. The Chalk of north Norfolk extends from the Cenomanian to the Lower Maastrichtian (Burke *et al.* 2010, fig. 1); the latter is unusually young for the English Cretaceous succession (Chatwin, 1961, p. 35; Peake & Hancock, 1961; Rawson *et al.*, 1978, pp. 30, 52; Moorlock *et al.*, 2002, pp. 3–5). The specimens are Late Cretaceous in age and older than the Late Maastrichtian (circa 70 million years old or more). Rafts of Chalk thrust by glacial ice during the Pleistocene are similarly Campanian–Maastrichtian based on both micro- and macrofossil evidence (Burke *et al.* 2010, pp. 621–623).'

The high ground in the field area is the Cromer Ridge, an east–west structure between Holt and Trimingham (Holt-Wilson, 2011, p. 18). Its complex origin was during Pleistocene glaciations, when glaciers 'bulldozed' superficial deposits to form the Ridge. Also included were rafts of Chalk which were intruded into these younger deposits, some of which are seen between Overstrand and Cromer. The superficial deposits are prone to landslides, and I recommend treating these potentially unstable cliffs with the respect that they deserve.

The beach at Overstrand is about [NGR TH 249 410]; that at Cromer, east of the pier, is about [NGR TH 227 420] (Fig. 47.1). Although dominantly sandy, the beach also has numerous pebbles and cobbles, the majority of which are locally derived from the Upper Cretaceous, including cobbles of flint and, less commonly, Chalk. I suspect these came mainly from offshore rather than from the cliffs.

What to look for

I commonly aim to be in north Norfolk in mid-March to mid-April. Timing is everything and I plan my trips to exploit, hopefully, any interesting clasts washed up on the beach by winter storms. But it will still be windy, cold and possibly wet; you should be well wrapped up in this field.

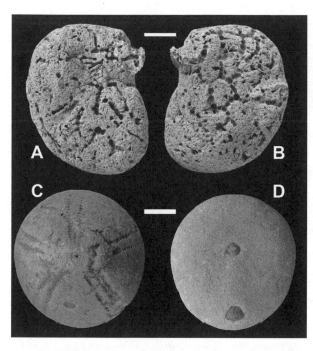

Figure 47.2 Chalk fossils from the north Norfolk coast between Overstrand and Cromer. (**A, B**) Two surfaces of a kidney-shaped Chalk pebble from the beach at Overstrand, Naturalis Biodiversity Center, Leiden, RGM 544 413 (after Donovan & Lewis, 2010, fig. 1). Look carefully and you will see the belemnite in section (top right in **A**, top left in **B**). The clast and belemnite guard have been bored by Recent sponges, *Entobia* isp. cf. *E. laquea* Bromley & d'Alessandro. (**C, D**) Irregular echinoid *Galerites* sp., RGM 621 012 (after Donovan, 2012, fig. 3A, E). (**C**) Apical view. (**D**) Oral view; peristome (centre) and periproct (posterior) occluded by flint. Specimens uncoated. All scale bars represent 10 mm.

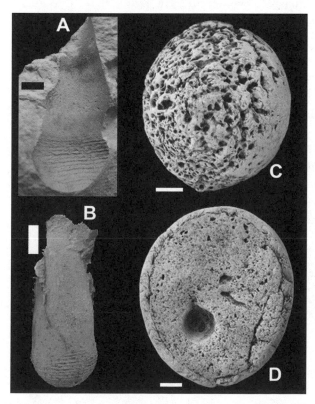

Figure 47.3 Recent borings from the north Norfolk coast between Overstrand and Cromer. (**A, B**) *Gastrochaenolites ornatus* Kelly & Bromley, RGM 617 814, in Chalk (after Donovan, 2011a, figs 2, 3, respectively). (**A**) Part of the boring, broken into three pieces by a person unknown. (**B**) Latex cast taken from the restored specimen. Coated with ammonium chloride. (**C, D**) *Echinocorys* ex gr. *E. scutata* Leske (after Donovan & Lewis, 2011, fig. 3a, c). (**C**) RGM 617 803, apical view, densely infested with Recent *Entobia* isp. (**D**) RGM 617 804, the apical surface has been largely abraded away; the most prominent boring is the base of *Gastrochaenolites* isp. Specimens uncoated unless stated otherwise. All scale bars represent 10 mm.

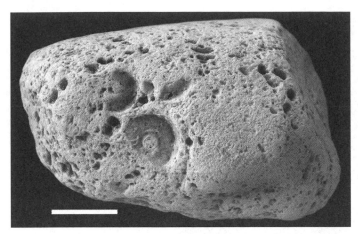

Figure 47.4 An erratic from the north Norfolk coast between Overstrand and Cromer (after Donovan, 2010, pl. 3). Derbyshire screwstone (Mississippian), RGM 544 455, with external moulds of two large crinoid columnals, *Megistocrinus? globosus?* (Phillips). Coated with ammonium chloride. Scale bar represents 10 mm.

Chalk fossils: Fossils in these Cretaceous clasts include rare sponges, inoceramids, belemnites (Donovan & Lewis, 2010) and echinoderms (Donovan & Lewis, 2011; Donovan, 2012, 2013) (Fig. 47.2). My own collecting has focused on the echinoids. The most common echinoid is the irregular *Echinocorys* Leske (Donovan & Lewis, 2011; Fig. 47.3C, D), preserved as calcitic tests and flint *steinkerns* (= internal moulds) and external moulds. Other echinoid species are few and rare (Donovan, 2012; Fig. 47.2C, D). A single crinoid columnal from the Chalk, *Austinocrinus bicoronatus* (von Hagenow), is about the size of a shirt button (Fig. 14.1).

Recent borings: Chalk clasts and bioclasts are commonly bored (Donovan 2011a, b, 2017; Donovan *et al.*, 2019) (Fig. 47.3). Most obvious are the large borings of bivalves, *Gastrochaenolites* isp. (Fig. 47.3A, B, D). The most interesting specimen was a belemnite, bored by sponges (*Entobia?* isp.) in the Cretaceous and again bored by sponges in the Recent (Fig. 47.2A, B).

Erratics: Glacial and fluvial erratics on Norfolk's beaches are dominantly lithics (http://www.northfolk.org.uk/geology/erratics.html). Fossiliferous erratics are rare, but include a crinoid columnal in a Derbyshire screwstone (Mississippian; Donovan, 2010; Fig. 47.4 herein) and vertical burrows of possible Mississippian age (Donovan, 2011b, fig. 3).

When you get to Cromer there are other things to do. Lunch in one of the fish restaurants is a must. There were still two second-hand bookshops last time that I looked, and both have natural history sections. The Cromer Museum is an essential place to visit, with a fine display of local geology and fossils supported by a small, but well-stocked gift shop, including books.

References

Burke, H., Phillips, E., Lee, J.R. & Wilkinson, I.P. (2010) Imbricate thrust stack model for the formation of glaciotectonic rafts: an example from the Middle Pleistocene of north Norfolk, UK. *Boreas*, **38**: 620–637.

Chatwin, C.P. (1961) *British Regional Geology. East Anglia and Adjoining Areas.* Fourth edition. Her Majesty's Stationery Office, London.

Donovan, S.K. (2010) A Derbyshire screwstone (Mississippian) from the beach at Overstrand, Norfolk, eastern England. *Scripta Geologica Special Issue*, **7**: 43–52.

Donovan, S.K. (2011a) The Recent boring *Gastrochaenolites ornatus* Kelly & Bromley, 1984, in a Chalk cobble from Cromer, England. *Bulletin of the Mizunami Fossil Museum*, **37**: 185–188.

Donovan, S.K. (2011b) Aspects of ichnology of Chalk and sandstone clasts from the beach at Overstrand, north Norfolk. *Bulletin of the Geological Society of Norfolk*, **60** (for 2010): 37–45.

Donovan, S.K. (2012) Taphonomy and significance of rare Chalk (Late Cretaceous) echinoderms preserved as beach clasts, north Norfolk, UK. *Proceedings of the Yorkshire Geological Society*, **59**: 109–113.

Donovan, S.K. (2013) Curiouser and curiouser: more on reworked *Echinocorys* (Echinoidea; Late Cretaceous) on the beaches of north Norfolk, eastern England. *Swiss Journal of Palaeontology*, **132**: 1–4.

Donovan, S.K. (2017) Neoichnology of Chalk cobbles from north Norfolk, England: implications for taphonomy and palaeoecology. *Proceedings of the Geologists' Association*, **128**: 558–563.

Donovan, S.K. with Donovan, P.H. & Donovan, M. (2019) A recurrent trinity of Recent borings in clasts around the southern and western North Sea. *Bulletin of the Geological Society of Norfolk*, **68** (for 2018): 51–63.

Donovan, S.K. & Lewis, D.N. (2010) Notes on a Chalk pebble from Overstrand: ancient and modern sponge borings meet on a Norfolk beach. *Bulletin of the Geological Society of Norfolk*, **59** (for 2009): 3–9.

Donovan, S.K. & Lewis, D.N. (2011) Strange taphonomy: Late Cretaceous *Echinocorys* (Echinoidea) as a hard substrate in a modern shallow marine environment. *Swiss Journal of Palaeontology*, **130**: 43–51.

Holt-Wilson, T. (2011) *Geological Landscapes of the Norfolk Coast: Introducing five areas of striking geodiversity in the Norfolk Coast Area of Outstanding Natural Beauty.* Norfolk Coast Partnership, Fakenham.

Moorlock, B.S.P., Hamblin, R.J.O., Booth, S.J., Kessler, H., Woods, M.A. & Hobbs, P.R.N. (2002) *Geology of the Cromer district – a brief explanation of the geological map Sheet 131 Cromer.* British Geological Survey, Keyworth.

Peake, N.B. & Hancock, J.M. (1961) The Upper Cretaceous of Norfolk. In: Larwood, G.P. & Funnell, B.M. (eds) *The Geology of Norfolk. Transactions of the Norfolk and Norwich Naturalists' Society*, **19** (6): 293–339.

Rawson, P.F., Curry, D., Dilley, F.C., Hancock, J.M., Kennedy, W.J., Neale, J.W., Wood, C.J. & Worssam, B.C. (1978) *A correlation of the Cretaceous rocks of the British Isles.* Geological Society, London, Special Report, **9**: 1–70.

CHAPTER 48

FIELD TRIP: CLEVELEYS, LANCASHIRE

Preamble

I did not open *The Times* until lunchtime, in a café out of the rain and wind. It would not have stopped me, but I would have been better informed. On page 3 (Friday, 16 August 2019), a headline clearly stated 'Month's worth of rain to fall today'. It was a warning read too late. The whole point of beachcombing at Cleveleys today was to see it in the sun; all previous visits were in December. More fool me: I was wet.

How to get there

Take the train from Manchester Oxford Road to Blackpool North via Bolton and Preston. At Blackpool North you should walk down the hill to the seafront and take a northbound tram (destination Fleetwood Ferry), leaving it at Cleveleys. Alight from the tram and turn left, back the way you just came, for a short distance and turn right (west) into Victoria Road West for the seafront, where you turn right (north) and examine the myriad cobbles on the beach as you walk (Fig. 48.1).

Geological history

In better weather than I suffered above, the mountains of the Lake District are starkly visible to the north, across the expanse of Morecambe Bay (Fig. 48.2). What now forms a major upland barrier to man was not an impediment to Pleistocene glaciers, originating in the north, including in the Lake District itself, and moving south (Smith, 2008). Glaciers pluck rocks at their base and sides, transporting them as a frozen bedload, rounding them and grinding them down. When the glaciers retreated, these transported clasts, ranging in grain size from muds to boulders, were left behind as an unsorted geological Irish stew once called boulder clay, a most descriptive term. As the boulder clay cliffs of the Lancashire coast were eroded away, the larger clasts were less easy to transport offshore. At Rossall Beach at Cleveleys, there is a superb array of diverse clasts,

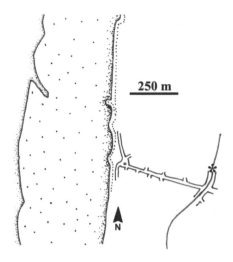

[Left] Figure 48.1 Outline map of the Cleveleys area. Alight from the tram at the Cleveleys stop (*) and walk west to the coast, then turn north along the beach (coarse stipple). Most clasts accumulate at the top of the beach, including adjacent to the seawall (dense stipple). The low tide mark is shown as a fine stipple; the dotted line is the footpath at the top of the seawall.

[Below] Figure 48.2 A view north from high on the beach at Cleveleys, Fylde, north-west Lancashire (after Donovan, 2020, fig.1). The outline of the peaks of the Lake District is plainly visible; it is a sunny, near-windless day in late December and the Irish Sea is particularly calm. The cobbles on the beach are obvious and diverse, a marvellous place for geological beachcombing.

essentially the geology of the Lake District (and, probably, elsewhere), but as a series of pebbles and cobbles on the seashore (Fig. 48.2; Ellis, 1968, p. 144).

What to look for

There are no numbered localities in this field guide (Fig. 48.1). The clasts on the beach, mainly pebbles and cobbles, are in constant motion at high tide and in storms. I look for promising specimens from my full height, but soon weary of bending down to pick up likely clasts. So, I invariably take a sturdy walking stick, which helps me to push upright without taking too much out of my back and leg muscles. There is usually at least a light breeze and the possibility of rain, so dress accordingly and have reserves

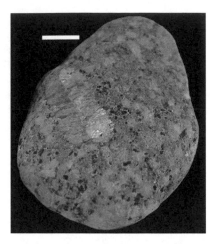

Figure 48.3 A pebble of the beautiful Shap Granite (collection of the Naturalis Biodiversity Center, Leiden), collected from the beach at Cleveleys, Fylde, north-west Lancashire. Note the large, twinned, pink phenocrysts of alkali feldspar (towards 10 o'clock). Scale bar represents 10 mm.

(anorak, sweater) in your field bag. My first line of defence against the elements – sun, rain, wind – is a broad-brimmed hat, always.

This is a beach for any geologist because the Lake District has supplied such a plethora of rock types. As a diversion, I have been fascinated to meet again the Shap Granite and its twinned alkali feldspar phenocrysts (Fig. 48.3). I have not been to the Shap pink quarry since 1982 or so, but recommend it as a site deserving of a visit (Firman *in* Moseley, 1978, pp. 156–158).

Back to palaeontology. Fossiliferous rocks on Rossall Beach, Cleveleys, are our old friend, the Carboniferous limestone (Mississippian) which crops out as an incomplete 'ring' on three sides of the Lake District dome (Murphy, 2015). The fossiliferous clasts are best seen when they are wet, which emphasizes the contrast between fossils (white) and sedimentary rock (grey). Typical Mississippian fossils include disarticulated crinoid ossicles, brachiopods, bryozoans and corals. As with building stones, there is the problem of identifying a three-dimensional fossil in two dimensions. Unlike the commonly planar surfaces of buildings, however, the curved surfaces of rounded clasts provide different aspects of a fossil that is easy to collect.

The colonial corals of these clasts are particularly enchanting. I have identified two genera, but there are most likely more. The tabulate *Syringopora* sp. (Fig. 48.4A) has fine branches and the skeleton is tubular with tabulae, but no septa. The rugosan *Lithostrotion* sp. (Fig. 48.4B) is similarly divided, but more massive, with branches of greater diameter. Internally the corallite has radial septa and concentric dissepiments that are particularly obvious near the circumference of each branch.

The other interesting features of the limestone clasts are modern borings. Although not trace fossils, they are modern traces. Three boring ichnogenera are known from this part of the coast (Donovan, 2020). Two of these are common, namely the U-shaped borings of polychaete worms, *Caulostrepsis taeniola* Clarke, and the labyrinthic networks of clionaid sponges, *Entobia* isp. (Fig. 48.5A, B, respectively). Much rarer are the club-shaped borings of bivalve molluscs, *Gastrochaenolites turbinatus* Kelly and Bromley (Fig. 48.5C).

References

Donovan, S.K. (2020). Recent borings in glacial erratics (Carboniferous Limestone), Cleveleys, Lancashire. *The North-West Geologist*, **21**: 31-49

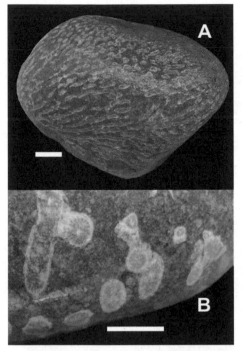

[Left] Figure 48.4 Mississippian colonial corals, Rossall Beach, Cleveleys, north-west Lancashire (both collection of the Naturalis Biodiversity Center, Leiden). (A) Cobble with *Syringopora* sp. (B) Details of *Lithostrotion* sp. showing radial septa and branching. Both clasts made wet for photography. Scale bars represent 10 mm.

[Below] Figure 48.5 Modern borings in Mississippian limestone clasts, Rossall Beach, Cleveleys, north-west Lancashire (after Donovan, 2020, figs 2F, 3D, F, 5A, C, respectively). All specimens in the collection of the Naturalis Biodiversity Center, Leiden (prefix RGM). (A) RGM.1332351, *Caulostrepsis* sp. cf. *C. taeniola* Clark. Borings are in the limestone, but not in the calcite veins, such as top centre to 4 o'clock. (B, C) *Entobia* isp. (B) RGM.1332357, well-exposed, colonial boring system, the outermost few mm of limestone having been lost through corrasion; more mature borings towards top. (C) RGM.1332360, mainly apertures of sponge colony. (D, E) *Gastrochaenolites turbinatus* Kelly and Bromley. (D) RGM.1332369, surface with shallow, basal remnants of bivalve borings. (E) RGM.1332370, a latex mould of the most complete boring of *G. turbinatus*; the mould is coated with ammonium chloride. Scale bars represent 10 mm.

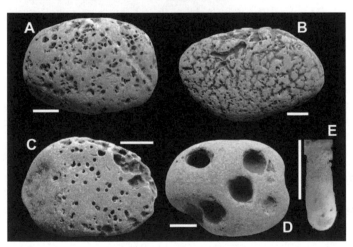

Ellis, C. (1968) [first published 1954]. *The Pebbles on the Beach*. Faber & Faber, London.

Moseley, F. (ed.) (1978) *The Geology of the Lake District*. Yorkshire Geological Society, Yorkshire.

Murphy, P. (2015) *Exploring the Limestone Landscapes of the Cumbrian Ring. BCRA Cave Studies Series 20*. British Cave Research Association, Buxton.

Smith, A. (2008) *The Ice Age in the Lake District. The Landscapes of Cumbria No. 3*. Rigg Side Publications, Keswick.

CHAPTER 49

FIELD TRIP: QUEEN VICTORIA'S BATHING BEACH, ISLE OF WIGHT

Preamble

This is a slightly different sort of field guide. Family holidays have a habit of going where the family wants to go and what better place to go than the Isle of Wight? It has all the ingredients of family fun, plus things that I want to do, such as browse in second-hand bookshops, good walking and riding on the IoW Steam Railway. Geology looms large in the island, but have a care: when I first visited the island in 1999 with my young family, my late wife assumed that I would be geologizing at some point and was surprised when I said a particularly firm 'no'. The Isle of Wight is one of those places of intense geological interest where every square metre is claimed by some expert researching something or other. I like a quiet life, and fighting over priority was and is not important.

When I returned with my older and larger family in 2012, I was, perhaps, more belligerent, but also had spread my wings to include such interests as building stones and modern traces, not primary concerns of any of the Isle of Wight geologists. I combined these for my first publication on the island's geology in which I described modern marine borings in cobbles that had been used to fix a wall around a church (Donovan, 2013). Further holidays on the island have involved yet more eccentric geological investigations with the approval of my family, bless 'em, mainly ichnological (Donovan & Isted, 2016), but also using the closed railway lines as a tool for fieldwork (Donovan, 2015).

This one-stop field guide integrates your fieldwork with family fun. Set up 'base camp' with some deckchairs in the shade, have an ice cream and a coffee, and relax with a good book. When the urge for fieldwork comes over you, walk along the beach and collect.

How to get there

Osborne House, formerly the country retreat of Queen Victoria and Prince Albert, is just outside East Cowes, Isle of Wight (Fig. 49.1). Several bus routes pass the front gate. Wherever you are in the Isle, there will be a bus to Newport, where you can change onto a bus to East Cowes, but be ready to alight before the terminus. There is ample space in the car park. The house opens at 10.00 a.m. There is a minibus service from the house to Queen Victoria's bathing beach, although it is also a pleasant walk. I recommend the deckchairs on the grass and under the trees, near the refreshment pavilion. Your family will love the setting, overlooking the Solent with the mainland in the distance.

Geological history

There are many volumes on the geomorphology, geology and palaeontology of the Isle of Wight; two favourites that I recommend from my own library are White (1921) and Insole *et al.* (1998). In the East Cowes area the succession is Oligocene and Quaternary (British Geological Survey, 2013). Rocks are not exposed *in situ* on the beach, but clasts derived from thin, fossiliferous beds are not uncommon. Where present, the fossils are abundant, particularly bivalve molluscs.

What to look for

Queen Victoria's private bathing beach at Osborne House, East Cowes [NGR 525 953] (Figs 49.1, 49.2), is rich in disarticulated oyster valves, other shells and limestone pebbles, the latter likely derived from the Paleogene (Insole *et al.*, 1998, pp. 18–22; British Geological Survey, 2013). See Lloyd & Pevsner (2006, figure on p. 211) for an estate map; the beach is in the area of locality 7 ('Landing House') therein.

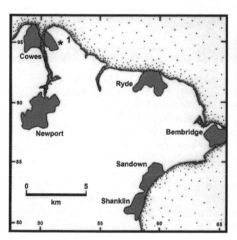 

[Left] Figure 49.1 Outline map of the eastern and northern Isle of Wight (modified after Donovan, 2014, fig. 1). Coastline and sea stippled; principal towns grey. Key: * = Osborne House; 1 = Queen Victoria's private beach at Osborne House.

[Right] Figure 49.2 Queen Victoria's private beach at Osborne House, July 2013, looking towards East Cowes. The beach is rich in reworked clasts of Palaeogene limestone and Recent shells with instructive borings.

Figure 49.3 The polychaete worm boring *Caulostrepsis taeniola* Clarke in limestone lithoclasts (**A, E**) and disarticulated oyster valves from Queen Victoria's bathing beach, Isle of Wight (after Donovan, 2017, fig. 1). Specimens in Figures 49.3 to 49.5 are all deposited in the Naturalis Biodiversity Center, Leiden (prefix RGM). (**A**) RGM 792 577, form 1, low-domed, densely infested surface of limestone clast. The reverse surface is less densely infested and gently concave; the lithoclast was probably turned in stormy conditions, thus exposing different surfaces to infestation. Note the bivalve boring *Gastrochaenolites* isp. at top, right of centre. Scale bar represents 50 mm. (**B**) RGM 792 579, form 1, moderately densely infested. One poor *Caulostrepsis* isp. occurs on the inner surface. (**C**) RGM 792 580, form 2, straight specimen partly exposed on inner surface of oyster valve. (**D**) RGM 792 581, form 2, curved specimen well exposed on inner surface of oyster valve. (**E**) RGM 792 578, form 1, flat lithoclast, densely infested on both this and the lower surface (not figured). Scale bar represents 50 mm. (**F**) RGM 792 582, form 2, sinuous specimen partly exposed on outer surface of oyster valve. All scale bars represent 10 mm unless stated otherwise.

I leave the study of these limestone clasts to you (Fig. 49.3A, E); I have been enthralled by the modern traces on this beach, which have provided unique demonstrations of ecology that is applicable to the fossil record. The modern ichnofauna includes common *Caulostrepsis taeniola* Clarke, *Caulostrepsis* isp., *Entobia* isp. and *Oichnus simplex*, with rare *Trypanites*? isp. and seagull beak marks (Donovan, 2014). This simple list hides some subtle features. For example, the U-shaped boring *Caulostrepsis taeniola* Clarke is common in both limestone clasts and oyster valves (Fig. 49.3). It falls into two groups, based on size. Members of Form 1 are small and gregarious, particularly in limestone. Form 2 is comprised of significantly larger individuals that are solitary, commonly infesting the disarticulated valves of dead oysters. Such borings are most commonly produced by the spionid polychaete worm *Polydora* spp. (Bromley, 2004, p. 460). These borings are considered to represent the spoor of at least two species, one small and the other larger. But remember, it is possible for two or more species to produce identical trace fossils (Donovan, 2017).

Small round holes in shells belong to the ichnogenus *Oichnus* Bromley and are commonly the spoor of predatory gastropods. Multiple oyster valves on Queen Victoria's bathing beach, *Ostrea edulis* Linnaeus, contain sparse *Oichnus simplex* Bromley, commonly infesting the free (right) valve. Borings are of consistent diameter, are commonly limited to one boring per valve, and may be penetrative (complete) or non-penetrative (incomplete) (Fig. 49.4). These borings show no preferred site; rather, they are as likely to penetrate the substrate marginally (thin valve) as centrally (thick valve). One borehole preserves a small bivalve, either a borer or nestler. It is suggested that these were not the products of predatory gastropods. Rather, they are bivalve borings into a pavement of dead oyster valves on a firm substrate. The small diameter of borings indicates that they represent an early, post-larval infestation. If the oyster valves were removed soon after by a storm, they would appear to have been bored pre-mortem by carnivorous snails, rather than post-mortem by bivalves constructing domiciles (Donovan, 2019).

Limpets are a group of gastropods that generate distinctive traces – their homing scars – and whose shells are bored by both predators and invertebrates forming domiciles. A Recent shell of *Patella vulgata* Linnaeus from Queen Victoria's bathing beach shows a distinctive and unusual preservation. The specimen preserves only the lower third of the shell, adjacent to the aperture (Fig. 24.4). The surface of breakage is densely infested by the U-shaped boring *Caulostrepsis taeniola* Clarke, which we have met above. The inferred taphonomic history is that the lower third of the dead shell was protected by burial in sediment while the apex was destroyed by a dense infestation of boring spionids. Such preservation is highly distinctive and might not be recognized in the fossil record when shells are only seen in two dimensions in a lithified rock surface (Donovan, in press).

One last comment. You will notice, glimpsing through the references, that most of my papers on the ichnology of the Isle were published in *Wight Studies: Proceedings of the Isle of Wight Natural History & Archaeological Society*. (It is a bit of a mouthful, but I refer

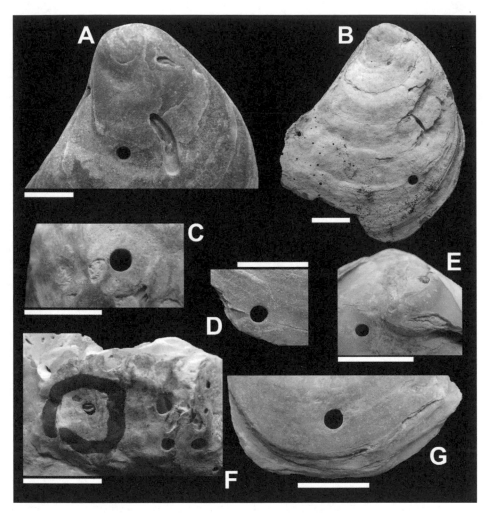

Figure 49.4 *Oichnus simplex* Bromley in *Ostrea edulis* Linnaeus from Queen Victoria's bathing beach, Isle of Wight (after Donovan, 2019, fig. 1). (**A**) RGM.1332386, slightly conical boring through moderately thick free valve; *Caulostrepsis taeniola* Clarke form 2 to right. (**B**) RGM.1332400, cylindrical boring through thin valve; *Entobia* isp. (sponge borings) to lower left. (**C**) RGM.1332389, cylindrical boring through moderately thick part and close to centre of valve. (**D**) RGM.1332387, cylindrical boring close to commissure. (**E**) RGM.1332388[3], slightly conical boring through thick part of valve near umbo. (**F**) RGM.1332396, a cylindrical *O. simplex* with an *in situ* bivalve borer(?) or nestler(?) and in a valve otherwise riddled by *Entobia* isp. (**G**) RGM.1332392, an incomplete boring in a thick part of a valve, near the adductor muscle scar and close to the commissure. All scale bars represent 10 mm.

to it as the *Proceedings* for brevity.) Back issues of the *Proceedings*, found in second-hand bookshops around the island, do have some geological papers, although archaeology and field biology dominate. By publishing in the *Proceedings*, I am putting something back for the community of local naturalists to read. This is part of a two-way process.

References

British Geological Survey. (2013) *Isle of Wight. England and Wales Special Sheet incorporating parts of sheets 330, 331, 344 and 345. Bedrock and Superficial Deposits. 1:50,000*. British Geological Survey, Keyworth, Nottingham.

Bromley, R.G. (2004) A stratigraphy of marine bioerosion. *In*: McIlroy, D. (ed.) *The Application of Ichnology to Palaeoenvironmental and Stratigraphic Analysis*. Geological Society, London, Special Publication **220**, 455–479.

Donovan, S.K. (2013) Neoichnology of the parish church of All Saints, Freshwater, Isle of Wight. *Wight Studies: Proceedings of the Isle of Wight Natural History & Archaeological Society*, **27**: 70–75.

Donovan, S.K. (2014) Bored oysters and other organism–substrate interactions on two beaches on the Isle of Wight. *Wight Studies: Proceedings of the Isle of Wight Natural History & Archaeological Society*, **28**: 59–74.

Donovan, S.K. (2015) Geological rambles on closed railways in the Isle of Wight. *Wight Studies: Proceedings of the Isle of Wight Natural History & Archaeological Society*, **29**: 81–88.

Donovan, S.K. (2017) Two forms of the boring *Caulostrepsis taeniola* Clarke on Queen Victoria's bathing beach, East Cowes. *Wight Studies: Proceedings of the Isle of Wight Natural History & Archaeological Society*, **31**: 98–101.

Donovan, S.K. (2019) Round holes in oysters, Queen Victoria's bathing beach, Osborne House, Isle of Wight. *Wight Studies: Proceedings of the Isle of Wight Natural History & Archaeological Society*, **33**: 92–96.

Donovan, S.K. (in press). Taphonomy of a limpet. *Ichnos*.

Donovan, S.K. & Isted, J. (2016) Borings ancient and modern: reworked oysters from the Ferruginous Sands Formation (Lower Cretaceous) of the Isle of Wight. *Wight Studies: Proceedings of the Isle of Wight Natural History & Archaeological Society*, **30**: 153—157.

Insole, A., Daley, B. & Gale, A. (1998) *The Isle of Wight. Geologists' Association Guide*, **60**: v+132 pp.

Lloyd, D.W. & Pevsner, N. (2006) *The Buildings of England. The Isle of Wight*. Yale University Press, New Haven.

White, H.J.O. (1921) [third impression 1975]. *A short account of the Geology of the Isle of Wight. Memoirs of the Geological Survey of Great Britain: England and Wales*. HMSO, London.

CHAPTER 50

FIELD TRIP: SALTHILL QUARRY, CLITHEROE, LANCASHIRE

Preamble

The locality at Salthill Quarry in Clitheroe that I describe below is my favourite fossil site in northern Europe (Fig. 50.1). It has deteriorated over the years – although a Site of Special Scientific Interest for its geology and palaeontology, it is under the management of a wildlife trust that puts its main effort into maintaining the flora (Donovan, 2011; Donovan *et al.*, 2014). However, even in such a poor state, it is still possible to collect a diversity of Mississippian fossils, dominantly crinoids, but also groups such as rarer rugose and tabulate corals, brachiopods, bryozoans and echinoids. There are several published field guides to Salthill Quarry (such as Grayson, 1981; Kabrna, 2011) and I have published guides to the fossil echinoderms (including Donovan, 1992; Donovan & Lewis, 2011). The present excursion is selective and you will want to read at least some of these more inclusive guides. A useful Ordnance Survey topographic sheet is OL41, 1:25,000 sheet *Forest of Bowland & Ribblesdale*.

How to get there

Locality 50.1: I assume that you will travel by rail. Trains leave Manchester Victoria for Clitheroe about once per hour (two per hour in rush hour). The train will terminate at Clitheroe on platform 2. Leave the platform; right down the slope; right under the bridge; right up the other side and past the station building (trains to Manchester leave from this side). Turn left away from the railway; walk past The Station pub and police station (both left), and the Old Post House Hotel (right, currently closed), and gently uphill in King Street. At the top of the hill turn right into Castle Street. Clitheroe Castle will become big and obvious on the right. Walk up the slope to the Castle Museum entrance [about NGR SD 7425 4160]. After viewing the exhibits, take a break at the Atrium Café.

Locality 50.2: Walk back down Castle Street and past the junction of King Street, through Market Place and down the hill of York Street. Turn right at the roundabout onto Waterloo Road; after about 200 m, turn left at the Royal Oak pub into Salthill Road. The end of Salthill Road is a cul-de-sac, going uphill. This deteriorates into a track of

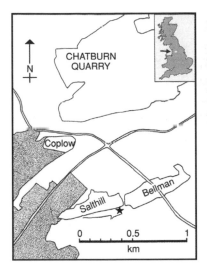

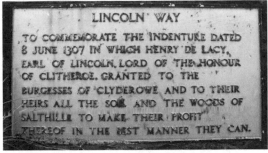

[Left] Figure 50.1 Map of the north-east Clitheroe area, showing the relative geographical positions of Coplow and Salthill (both disused), and the working Bellman and Chatburn quarries (after Donovan, 1992, fig. 1). The star indicates the position of the surface of Locality 50.2, the scraped surface at Salthill Quarry. The inset map indicates the position of the main map (arrowed) in mainland Britain.

[Right] Figure 50.2 The explanatory plaque in Lincoln Way, Clitheroe, the road through the industrial estate situated in Salthill Quarry. This plaque was funded by Mr. Stanley Westhead (1910–1986), a notable local collector, whose fine collection is now in the Natural History Museum, London (Donovan, 2012).

asphalt and stones which continues gently rising. Walk straight on past Salthill Barn and Salthill Cottage. Stop to read the explanatory signboard to the Salthill Quarry area, erected by the Wildlife Trust for Lancashire, Manchester and North Merseyside. We are aiming for Site 6 on this map (Fig. 50.1), but you will see that there are other geological sites and features to tempt your indulgence.

Take the track, then steps on the right and down into Lincoln Way (Fig. 50.2). Note the bedded limestones on the right. Cross Lincoln Way and up the steps on the other side. Walk east-north-east along the top of the ridge to Site 6 (= Locality 50.2) towards the end. Relax in the stone crinoid seat before collecting at the bottom of the slope (Fig. 50.3).

Figure 50.3 Part of Locality 50.2 in Salthill Quarry, Clitheroe, January 2020, with a beaming Mr. Paul Kabrna for scale. It may not look much, but this is the best site for collecting the cups and thecae of fossil crinoids that I know. The surface on which Paul is standing is rich in invertebrate fossil debris.

Geological history

The reefal limestones of the Clitheroe basin were laid down in a tropical, carbonate ramp setting flanking the Craven Basin (Miller & Grayson, 1972; Aitkenhead *et al.*, 2002). Carbonate ramps are carbonate platforms with a depositional slope of only a shallow gradient. The reefs themselves were mud mounds. The lime mud was produced by microbes and rapidly lithified, providing a hard substrate for a diversity of invertebrates.

What to look for

Locality 50.1: This is the Clitheroe Castle Museum (Ashworth, 2010). This first stop is made so that you can examine the small display of local fossils and get a preliminary insight into the local geology. Be sure to check the opening times: in January 2020 the museum did not open until 12.00 noon. Ashworth (2010, p. 6), when discussing the geology of the Clitheroe district, stated: 'This area is designated a Site of Special Scientific Interest (SSSI), a protected area, because of the sheer variety of geological material, and it is strictly forbidden to remove any rock.' No hammering makes sense, but collecting nothing is nonsensical. Who gains if scientifically important specimens are not collected and investigated? Certainly, Locality 50.2 would benefit from further excavation (Donovan *et al.*, 2014), not less. A geological site remains relevant by being worked over, not by fencing off the orchids.

Locality 50.2: This site is the highly fossiliferous scraped slope of Salthill Quarry, Clitheroe (Figs 50.1, 50.3) (approximate NGR [SD 755 425]). It is point 3 of Grayson (1981) and locality 4 of Kabrna (2011). The Salthill Cap Beds of Miller & Grayson (1972) are Dinantian (Mississippian, Lower Carboniferous), Visean, lower Chadian (George *et al.* 1976, table 2; N. J. Riley *in* Donovan & Sevastopulo 1985, p. 179; Riley, 1990).

Figure 50.3 may not appear too tempting to the uninitiated, but look carefully. You will soon note that you are standing on gravel dominated by fossil invertebrates, many hundreds of them. This includes an abundance of crinoid debris, but specimens are many and varied. I have illustrated a few of my favourites, mainly my favourite crinoids, but also tabulate corals and trace fossils (Figs 30.1, 50.4).

Fragments of crinoid stems are bountiful, yet Salthill Quarry is best known for its crinoid cups and thecae (that is, cups preserving the 'lid' of the tegmen; see Donovan & Lewis, 2011, for a glossary of crinoid terminology). I expect to find a few of these on every visit. Camerate crinoids had a plated tegmen and commonly resemble a golf ball (Fig. 50.4A). Yet for some camerate species, like *Platycrinites jameswrighti* (Fig. 50.4B), the tegmen is unknown; indeed, this taxon is a rarity (I haven't seen a new cup of this species since the mid-1980s), so keep your eyes peeled. Cladid crinoids also had a robust cup, but the tegmen was less robust and commonly not preserved (Fig. 50.4C, E). Borings are not uncommon in crinoid pluricolumnals (Fig. 50.4F) and, more rarely, cups. And cherish slabs of bed such as Figure 50.4D, a piece of Mississippian seafloor and abounding with interesting invertebrates.

Figure 50.4 Some favourite fossils (mainly crinoids, I admit) from Locality 50.2; after Donovan & Lewis (2011) unless stated otherwise. All deposited in the Natural History Museum, London (BMNH), unless stated otherwise. (**A**) *Amphoracrinus turgidus* Wright, theca, BMNH E71261. (**B**) *Platycrinites jameswrighti* Donovan & Westhead, cup, BMNH E70771. (**C**) *Cyathocrinites planus* J.S. Miller, cup, BMNH E71151. (**D**) Slab of crinoidal biosparitic limestone, BMNH EE5717 (after Donovan, 2014, fig. 2)· A, two columnals of probable *Gilbertsocrinus* Phillips; B, pluricolumnal encrusted by tabulate coral colony, *Emmonsia parasitica* (Phillips); C, pluricolumnal with multiple borings and showing growth reaction; D, two specimens of the tabulate coral *Cladochonus* sp.; E, probable basal circlet of a crinoid cup; F, pluricolumnal of a platycrinitid crinoid; G, brachial ossicle (= plate of a crinoid arm); H, fragment of crinoid arm; I, indeterminate coral, showing beekite silicification. (**E**) *Poteriocrinites crassus* J.S. Miller, cup, BMNH E71393. (**F**) *Pentagonocyclicus*? (col.) sp., crinoid pluricolumnal, Naturalis Biodiversity Center, Leiden, RGM 791 810, infested by the boring *Oichnus paraboloides* Bromley (after Donovan & Tenny, 2015, fig. 2A). Scale bars represent 10 mm.

References

Aitkenhead, N., Barclay, W.J., Brandon, A., Chadwick, R.A., Chisholm, J.I., Cooper, A.H. & Johnson, E.W. (2002) *British Regional Geology: The Pennines and adjacent areas.* Fourth edition. British Geological Survey, Nottingham.

Ashworth, S. (2010) *Clitheroe Castle Museum: Castle Keep, Museum and Park.* Scala Publishers, London.

Donovan, S.K. (1992) A field guide to the fossil echinoderms of Coplow, Bellman and Salthill Quarries, Clitheroe, Lancashire. *North West Geologist*, **2**: 33–54.

Donovan, S.K. (2011) Salthill Quarry, Clitheroe: a resource degraded. *Deposits*, **25**: 46–47.

Donovan, S.K. (2012) Stanley Westhead and the Lower Carboniferous (Mississippian) crinoids of the Clitheroe area, Lancashire. *Proceedings of the Yorkshire Geological Society*, **59**: 15–20.

Donovan, S.K. (2014) Palaeoecology and taphonomy of a fossil 'sea floor', in the Carboniferous Limestone of northern England. *Mercian Geologist*, **18**: 171–174.

Donovan, S.K., Kabrna, P. & Donovan, P.H. (2014) Salthill Quarry: a resource being revitalized. *Deposits*, **40**: 32–33.

Donovan, S.K. & Lewis, D.N. (2011) Fossil echinoderms from the Mississippian (Lower Carboniferous) of the Clitheroe district. *In*: Kabrna, P.N. (ed.) *Carboniferous Geology: Bowland Fells to Pendle Hill*. Craven and Pendle Geological Society, Lancashire, 55–96.

Donovan, S.K. & Sevastopulo, G.D. (1985) Crinoid arms from Salthill Quarry, Clitheroe, Lancashire. *Proceedings of the Yorkshire Geological Society*, **45**: 179–182.

Donovan, S.K. & Tenny, A. (2015) A peculiar bored crinoid from Salthill Quarry, Clitheroe, Lancashire (Mississippian; Tournaisian), UK. *Proceedings of the Yorkshire Geological Society*, **60**: 289–292.

George, T.N., Johnson, G.A.L., Mitchell, M., Prentice, J.E., Ramsbottom, W.H.C., Sevastopulo, G.D. & Wilson, R.B. (1976) *A correlation of Dinantian rocks in the British Isles*. Geological Society Special Report, **7**: 1–87.

Grayson, R. (1981) *Salthill Quarry geology trail*. Nature Conservancy Council, London.

Kabrna, P. (2011) Excursion 5. Pendle Hill and Clitheroe. *In*: Kabrna, P. (ed.) *Carboniferous Geology: Bowland Fells to Pendle Hill*. Craven and Pendle Geological Society, Lancashire, 157–165.

Miller, J. & Grayson, R.F. (1972) Origin and structures of the Visean 'reef' limestones near Clitheroe, Lancashire. *Proceedings of the Yorkshire Geological Society*, **38**: 607–638.

Riley, N.J. (1990) Stratigraphy of the Worston Shale Group (Dinantian), Craven Basin, north-west England. *Proceedings of the Yorkshire Geological Society*, **48**: 163–187.

CHAPTER 51

FIELD TRIP: HURDLOW, DERBYSHIRE

Preamble

The Parsley Hay to Hurdlow field excursion combines three of my favourite interests, two of which are geological, namely Mississippian limestones, the landscape of the White Peak of Derbyshire and walking closed railways. The Buxton to Ashbourne and Buxton to Cromford lines, formerly of the London, Midland and Scottish Railway, were closed south of Dowlow by British Railways in the 1960s (Sprenger, 2013). The section walked in this excursion forms part of the long distance foot- and cycle-path, the Pennine Bridleway, from Dowlow to Cromford Wharf; the former railway junction for the Cromford and Ashbourne lines is south of Parsley Hay.

The surface of crushed limestone is good for walking and cycling. In what was formerly the goods yard at Parsley Hay, a visitor centre includes a shop, cycle hire, a café and public conveniences. Cycle hire includes disabled-friendly, three- and four-wheel electric carts, so there is a high probability that just about anyone can make this excursion, which is only about 6 km round trip (with a good lunch halfway!).

How to get there

For the motorist, aim for the Parsley Hay visitor centre [about NGR SK 147 637] off the A515 Buxton to Ashbourne road; it is somewhat closer to Buxton. There is a pay-and-display car park. The facilities listed above are open throughout the school summer holidays and at many other times (certainly not in mid-winter), but check in advance, particularly for cycle hire (Parsley Hay Cycle Hire, Parsley Hay near Buxton, Derbyshire SK17 0DG; telephone (01298) 84493; Email parsleyhay.cyclehire@peakdistrict.gov.uk).

For those of us who rely on public transport, it is not quite so easy. I travel by rail from Manchester Piccadilly to Buxton, where I walk down the slope opposite the station towards the town centre; there is a taxi rank towards the bottom of this slope and last time I was there, in January 2020, the fare was about £25 to Parsley Hay. The motorist will return to their car at Parsley Hay at the end of the day, but I tend to continue to Brierlow Bar, along the footpath on the south-west side of the major quarries about Hindlow. At

Brierlow Bar, the High Peak Bookstore offers a cup of tea, a slice of sponge cake and a browse of their eclectic collection. From here there are buses to Buxton; otherwise, a return taxi from Hurdlow or here is equally convenient.

Geological history

The Mississippian limestones of the White Peak of the Derbyshire Plateau were deposited between about 350 and 320 million years ago in a range of environments from shallower (including reefs) to deeper water. Coeval igneous action led to a range of basaltic lavas, intrusive sills and associated deposits, with mineralization related to the cherts that are locally interbedded with limestones and also the silicification of some shells. The limestones are highly fossiliferous, but massive. As the White Peak is part of a National Park, no hammering, but loose blocks in the undergrowth may include interesting specimens. Unlike the Carboniferous limestones at, for example, Salthill Quarry in Lancashire (Chapter 50), mudrock horizons are relatively few and it is these that may produce individual, collectible fossils (Donovan, 2018).

My interest in the White Peak of Derbyshire and Staffordshire is that it is one of the most crinoid-rich areas in the Carboniferous Limestone of the British Isles, yet is not widely recognized as such (Donovan, 2013). As I have already intimated, there are many thick sequences of limestone with locally common cherts and few fossiliferous mudrock horizons. Identifiable macrofossils in these well-lithified beds are brachiopods and corals. Crinoids in massive limestones may be abundant, but are mainly fragments of stem, at best difficult to name. For this reason, crinoids, despite being the commonest fossils in these rocks, have been largely ignored by systematic palaeontologists.

There are many guidebooks to the geology of the Peak District. Amongst the several on my bookshelf, I recommend Simpson (1982), Ford (1996), Cope (1999) and Broadhurst (2001), but this list is not exhaustive.

What to look for

Walk north from the Parsley Hay visitor centre and after about 200 m there is a circular, drystone-built building on the right (Fig. 51.1). It is a *kazun*, a shelter based on similar structures from the Istrian Peninsula of Croatia and elsewhere in Europe. The walls are limestone from the nearby Once-a-Week quarry and the 'false'-domed roof is sandstone from Johnson's Wellfield Quarry near Huddersfield. Stop and enjoy the fossils in the limestone slabs in the robust wall, which will be cleaner than any rock exposure on this expedition (if you want to see inside the structure, bring a torch). Crinoid pluricolumnals (lengths of stem) and productid brachiopods are common in section, along with rare solitary rugose corals.

The *kazun* was built after publication of Simpson's book; otherwise, we are following the route of the northern half of his excursion 14, his stops 5, 6 and 7 (1982, pp. 106–107,

Figure 51.1 *Kazun* at Parsley Hay. (**A**) The *kazun*, looking towards Parsley Hay; walking stick and backpack for scale (right). (**B**, **C**) Fossils in the walls (Carboniferous limestone) of the *kazun*. Scales in cm. (**B**) Oblique section through a solitary rugose coral, showing radial septa. (**C**) A long crinoid pluricolumnal.

Figure 51.2 From Parsley Hay to Hurdlow; Simpson's (1982) limestone localities. (**A**) The view north from the *kazun* – comfortable walking country. (**B**) Simpson's stop 5, now rather overgrown even in January. (**C**) Simpson's stop 7 with silicified productid brachiopods. (**D**) Simpson's stop 6 has deteriorated in almost 40 years since publication.

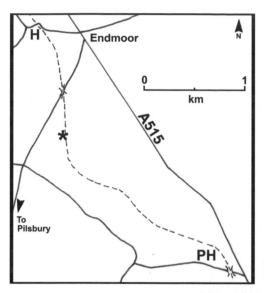

[Left] Figure 51.3 Locality map of the Parsley Hay–Hurdlow area, south-west of Buxton, Derbyshire. Key: solid lines = roads; dashed line = Pennine Bridleway; H = site of Hurdlow railway station (closed); PH = site of Parsley Hay railway station (closed), now a cycle hire centre and café; * = fossil locality discussed herein (= Simpson Stop 7). After Donovan & De Winter, (2019, fig. 2).

[Below] Figure 51.4 The Monsal Dale Limestones Formation (lower Brigantian, Visean), south of site of former Hurdlow railway station, Derbyshire (after Donovan & De Winter, 2019, fig. 1). (A, B) Productid brachiopods exposed in limestones on west side of cutting. Scale in cm. (C) View of the exposure looking south towards Parsley Hey.

figs 73, 74; fig. 70 therein is a locality map). Simpson's stops 5 and 6 are in limestones, but these sites have deteriorated in almost 40 years since the original publication (Fig. 51.2). Stop 5 [NGR SK 141 640] is a short railway cutting through the Bee Low Limestone Formation of the Peak Limestone Group. Fossils include 'shell and crinoid fragments …' (Simpson, 1982, p. 106) in a limestone showing patchy dolomitization; however, on the weathered faces, I have found no fossils. Simpson makes no mention of fossils at his stop 6 [NGR SK 132 646], another cutting, with a faulted contact between the younger Monsal Dale Limestone Formation and the older Bee Low Limestone Formation. There is an overbridge at the northern end of this site.

After two disappointing localities, Simpson's most northerly site, his stop 7, is of interest. It is exposed on the Pennine Bridleway south of the bridge over the minor road from Endmoor on the A515 south-west towards Pilsbury (Fig. 51.3). This site is a long, straight cutting centred at about [NGR SK 130 652] (Fig. 51.4C) in an area of typical limestone upland. *In situ* limestone beds are exposed on both sides of the cutting, but particularly on the west face. Specimens that I have collected came from float blocks of limestone in the grasses and other vegetation below this west face. This locality is in the Monsal Dale Limestone Formation.

No hammering, of course, but please examine the face and rummage through the loose blocks in the long grass. The face is locally packed with silicified productid brachiopods (Figs 51.2C, 51.4A, B) which also may be found in float blocks. Of particular interest on these float blocks are etched surfaces that reveal minute ossicles of crinoids (mainly columnals and pluricolumnals) and test plates of echinoids. From these, Donovan & De Winter (2019) identified two genera of echinoid and several morphologies of crinoid ossicle. Although preservation is indifferent, this is only the tenth echinoid site identified in the White Peak.

One last stop. Walk about ten minutes to the north, to the next overbridge, and make a detour to The Royal Oak pub, close to the site of the former Hurdlow station (H in Fig. 51.3). It is warm and dry, and the food is well worth taking a rest for. Relax and call your taxi, otherwise prepare to walk north to Brierlow Bar (see above) or back to your car at Parsley Hay.

References

Broadhurst, F.M. (2001) *Rocky Rambles in the Peak District: Geology beneath your Feet!* Sigma Leisure, Wilmslow, Cheshire.

Cope, F.W. (1999) *The Peak District*. Third revised edition. *Geologists' Association Guide*, **26**: i–iv + 1–78.

Donovan, S.K. (2013) Where are all the crinoids? An enigma of the Lower Carboniferous (Mississippian) White Peak of midland England. *Geology Today*, **29**: 108–112.

Donovan, S.K. (2018) Taphonomy of a Mississippian crinoid pluricolumnal, Newton Grange, Derbyshire, UK. *Proceedings of the Yorkshire Geological Society*, **62**: 59–63.

Donovan, S.K. & De Winter, A.J. (2019) Notes on Mississippian echinoderms from Hurdlow, Derbyshire, central England. *Proceedings of the Geologists' Association*, **130**: 582–589.

Ford, T.D. (1996) The Castleton area, Derbyshire. *Geologists' Association Guide*, **56**: i–iii + 1–93.

Simpson, I.M. (1982) *The Peak District*. Unwin Paperbacks, London.

Sprenger, H. (2013) *Rails to Ashbourne*. Kestrel Railway Books, Southampton.

CHAPTER 52

FIELD TRIP: ANTIGUA

Preamble

I spent twelve and a half happy and productive years at the University of the West Indies and have always been determined to include one Antillean field guide in this book, as a sort of 'what to do in your holidays' suggestion for palaeontologists whose significant other might prefer to sun themself on the beach. I have been to ten islands and territories in the Caribbean for fieldwork, and the choice for this guide has been tough. Logically, it should have been Jamaica, where I lived and know best, but a big field guide already exists (Donovan *et al.*, 1995). After much wringing of hands – Tobago and Barbados are also favourite field areas – I voted for Antigua.

How to get there

Catch a flight. From Europe British Airways or Virgin Atlantic, from North America take American Airlines or Air Canada, and their various possibilities within the Antillean region, such as LIAT.

Once in Antigua, it is best to hire a car. The island is small (Fig. 52.1), but air-conditioned comfort in the humid tropics will be appreciated. In 1993 I had a research grant and hired a taxi every day, which worked well, but a hire car is both cheaper and more flexible.

Geological history

(Adapted from Donovan *et al.*, 2014a.) Antigua is a small island, about 280 km², in the northern part of the Lesser Antilles volcanic arc. Antigua is one of the Limestone Caribbees, islands that were volcanic in origin in the mid-Cenozoic, but are now quiescent. Indeed, Antigua has a rock record that records the transition from island arc volcanism to quiescence and limestone deposition, all of which happened during the Late Oligocene (Fig. 52.1). The best exposures are found around the coast and in some inland quarries. Topographic maps include a useful 1:35,000 sheet, *Antigua and Barbuda*, published by International Travel Maps, 530, West Broadway, Vancouver,

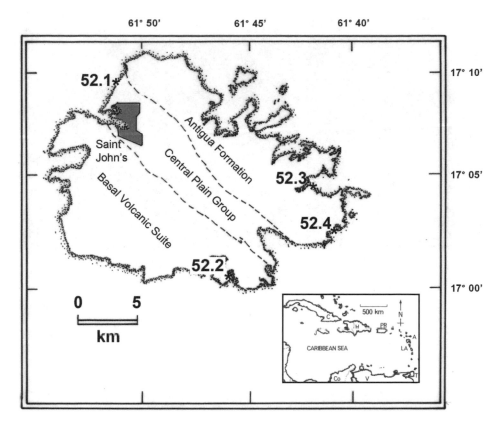

Figure 52.1 Outline map of Antigua, showing the principal geological subdivisions and the city of Saint John's. Inset map shows the position of Antigua in the Caribbean. Key (clockwise from Jamaica): J = Jamaica; C = Cuba; H = Hispaniola (Haiti+Dominican Republic); PR = Puerto Rico; A = Antigua (arrowed); LA = Lesser Antilles; T = Trinidad; V = Venezuela; Co = Colombia. Key to localities: 52.1 = CPG at Corbison Point; 52.2 = Nelson's Dockyard with fossil palms from the CPG in the Museum; 52.3 = Antigua Formation at Hughes Point; 52.4 = Antigua Formation at north Half Moon Bay (modified after Donovan et al., 2014a, figs 1, 2).

British Columbia, Canada V5Z 1E9. This is widely available; this and other maps can be purchased from Edward Stanford, 7 Mercer Lane, Covent Garden, London, WC2H 9FA.

Antigua has a mixed rock record of igneous overlain by sedimentary rocks. The regional dip of Antigua is towards the north-east (Fig. 52.1), the oldest (volcanic) rocks outcropping in the west and south. The rock record of the island is divided into three conformable units: the Basal Volcanic Suite (not discussed further); the Central Plain Group; and the Antigua Formation.

The Central Plain Group (CPG) outcrops in a low-lying belt extending from the north-west to south-east of the island (Fig. 52.1). The CPG consists of mixed siliciclastic and limestone sedimentary rocks of both marine and non-marine origin. The siliciclastic rocks were derived by weathering, erosion and re-deposition of the BVS. The most distinctive rocks of the CPG are silicified, including petrified wood, the national stone

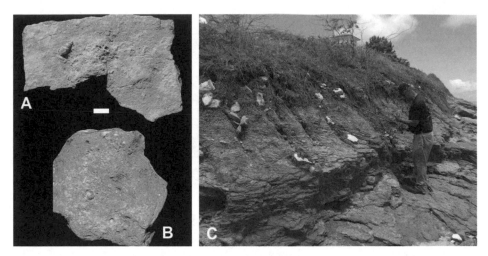

Figure 52.2 Central Plain Group (CPG) near Corbison Point, parish of Saint John, northern Antigua (Locality 52.1). (**A, B**) Hand specimens of chertified siliciclastic sedimentary rock bearing fossil freshwater snails, *Hemisinus* sp. in lateral view (**A**) and cross section (**B**, rounded structure slightly left and below centre) (specimens in collection of Naturalis Biodiversity Center, Leiden) (after Donovan *et al.*, 2014b, fig. 8). (**C**) The end of Corbison Point is private property, but on the north side, close to the beach, there are accessible exposures of CPG. The late Professor Trevor Jackson is examining a hand specimen of chert (after Donovan, 2014, fig. 7).

of Antigua (Locality 52.2), and cherts preserving abundant freshwater snails (Locality 52.1).

The Antigua Formation is a sequence of varied limestones with minor tuffaceous/sandy horizons that are exposed in the north and east of the island (Fig. 52.1). The differing limestone lithologies are indicative of contrasting palaeoenvironments of deposition; these have been differentiated in at least two Ph.D. theses, yet neither has been published. The macrofossils are varied and, in some sections, silicified. Common taxa include larger benthic foraminifers, scleractinian corals, benthic molluscs, echinoids, crabs and bryozoans (Locality 52.3). Deeper water environments high in the succession are indicated by the presence of deep water sponges, crinoids and brachiopods (Locality 52.4).

What to look for

Locality 52.1; Central Plains Group at Corbison Point: The CPG is low-lying and rarely well exposed. The best coastal site is north of St. John's at Corbison Point (Fig. 52.1, 52.2C). Although this now has limited access (whoever lives at the end likes big guard dogs), cherts with freshwater snails may be collected on the north side (Fig. 52.2A, B).

Locality 52.2; Museum at Nelson's Dockyard: Most famously, the CPG is known for its petrified (=silicified) wood, the national stone of Antigua and Barbuda, but this has been over-collected for hundreds of years. Impressive specimens of petrified wood are

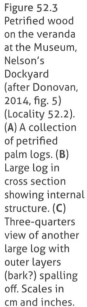

Figure 52.3 Petrified wood on the veranda at the Museum, Nelson's Dockyard (after Donovan, 2014, fig. 5) (Locality 52.2). (A) A collection of petrified palm logs. (B) Large log in cross section showing internal structure. (C) Three-quarters view of another large log with outer layers (bark?) spalling off. Scales in cm and inches.

on display at the Museum of Antigua and Barbuda in Saint John's, and in the Museum, Nelson's Dockyard.

Nelson's Dockyard is a handsome tourist spot on the south coast. Relax, have a beer and something to eat, but also visit the Museum. On the veranda is a fine display of logs of petrified wood from the CPG (Fig. 52.3). Anon (1818, p. 56) wrote a modern-sounding description: 'siliceous petrifactions of wood from Antigua. Their characters are indubitable; the distinct ligneous layers corresponding with annual growth; the medullary prolongations, the knots formed by branches, the cracks and the bark, are all distinctly visible. Some of the pieces are ponderous portions of large trees.' The mineralization was obvious to the trained eye: 'crystals of quartz are apparent in the cavities: some parts are agatized, and veins of chalcedony occasionally pervade the fissures ...' (pp. 56–57).

Petrification of wood requires silica, and the principal source on Antigua seems to have been the underlying volcanics. This process is permineralization; the silica is deposited directly on the organic structure of the wood before it decays and may lead to such fine fidelity of preservation that structures as delicate as the cell walls are identifiable.

Available identifications mainly classify these fossil logs as palms, including members of the form genus *Palmoxylon* Schenk. Identifications are based on fine structures apparent in petrologic thin sections, which rely on the high fidelity of preservation (Felix, 1883).

Locality 52.3; Antigua Formation at Hughes Point: The Hughes Point area is on the south coast of Nonsuch Bay, parish of St. Philip, eastern Antigua. Travel south-east and east through St Philip, admiring the views into Willoughby Bay to the south-west, and take the north turning through the western edge of Freetown. The road north of here is poor. Access is gained at the gates to the resort above Ayers creek. Explain that you intend to visit Mr. Rolf Ahlfors; the guard on the gate will direct you. Park at the end of the road, adjacent to the shore, and find Mr. Ahlfors at home, explaining your interest in the rocks at Hughes Point, and please pass on my best regards. Walk around the coast, wading in the shallow water to avoid the many fallen blocks.

Hughes Point has many advantages for the collector: moderately easy access; good exposures; and many large fallen blocks from high in the cliff. The upper and lower parts of the section represent different biofacies. Densely clustered trees obstruct the lower part of the section in places (these have all grown up since my first visit over 25 years ago), but the fallen blocks give access to the upper cliff face without climbing. Beds are well lithified and bedding is well defined, although horizons may have irregular bases and tops. Common fossils from the base of the section include the oyster *Hyotissa* and the echinoid *Clypeaster*, amongst others.

Fallen blocks are numerous and came from about 15–20 m above the shore. They are composed of straw-coloured, well-indurated calcarenites with rust-brown banding and small, spherical concretions. Some horizons contain abundant *Thalassinoides* isp. burrows, probably produced by mud shrimps. Tube-like accumulations of shelly debris are burrow fills (Donovan *et al.* 2017a). These boulders are richly fossiliferous, yielding common pectinid bivalves, spatangoid echinoid tests and asteroid marginal ossicles, with rarer bryozoans and branching corals.

Figure 52.4 Bored oyster, *Hyotissa antiguensis* (Brown) and cemented limestone, all derived from the Upper Oligocene Antigua Formation at Hughes Point, parish of St. Philip (Locality 52.3); all borings Recent (after Donovan *et al.*, 2014b, fig. 6). Specimens in the collection of the Naturalis Biodiversity Center, Leiden (prefix RGM). **(A)** RGM 791 217, *Rogerella*? isp. in centre, particularly the teardrop-shaped boring to the left; probably produced by acrothoracian barnacles. **(B)** RGM 791 230, outer surface of valve showing bio-eroded and corroded surface; bio-eroders *Entobia* isp. and *Gastrochaenolites* isp., produced by clionaid sponges and bivalves, respectively. **(C)** RGM 791 232, three valves cemented together, all highly bio-eroded by *Entobia* isp. **(D)** RGM 791 231, apertures of *Entobia* isp. on external surface of shell. Scale bars represent 10 mm.

The large Upper Oligocene oyster *Hyotissa antiguensis* (Brown) is locally common both *in situ* in extensive coastal exposure, and reworked as float on the beach and in shallow water, the latter associated with common bored clasts of limestone. Reworked valves and shells are large and robust, and have been infested by a range of modern boring and encrusting organisms (Donovan *et al.*, 2014b; Fig. 52.4 herein).

Locality 52.4; Antigua Formation at Half Moon Bay: The limestones exposed on the north-east side of Half Moon Bay (Figs 54.1, 54.5) are amongst the stratigraphically highest in the Antigua Formation and represent deeper water deposition than, for example, Hughes Point. Three fossil groups provide evidence for this deeper water environment,

Figure 52.5 View across Half Moon Bay, showing the extensive limestone exposure of the Antigua Formation and prominent sedimentary bedding. This is an idyllic spot to collect Oligocene shelly fossils.

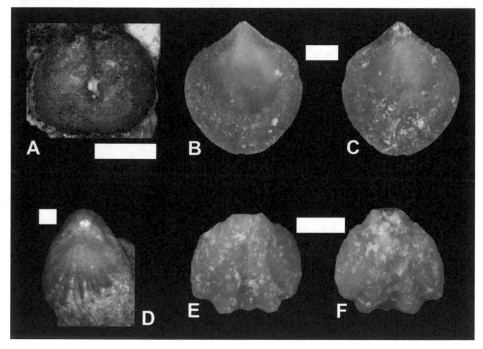

Figure 52.6 Brachiopods from the Antigua Formation (Upper Oligocene) (after Donovan *et al.*, 2015, fig. 4.). All specimens in the Florida Museum of Natural History, Gainesville (prefix UF). (**A**) *Cistellarcula* sp., UF 249932, Parham new quarry; width 10 mm. (**B, C**) Juvenile *Tichosina* sp., UF 239969, quarry south of naval base. (**D**) *Terebratulina* sp., UF 249993, Half Moon Bay (Locality 52.4); width 5 mm. (**E, F**) *Argyrotheca* sp., UF 239970, quarry south of naval base. Scale bars represent 1 mm unless stated otherwise. Specimens uncoated.

namely brachiopods, crinoids (columnals only) and large, thin-walled fossil sponges. Comparison with analogous modern environments indicates deposition in 150 + m water depth. That these groups are now recognized as occurring high in the formation implies deeper water deposition in what appears to be a retrograde succession (Donovan *et al.*, 2015, 2017b; Fig. 52.6 herein). Your significant other will enjoy the beach near the car park, where they can relax, swim, and buy cold drinks and hot food.

References

Anon. (1818) Petrified wood from Antigua. *American Journal of Science*, **1**: 56–57.

Donovan, S.K. (2014) The petrified wood of the Oligocene of Antigua, West Indies. *Deposits*, **40**: 12–14.

Donovan, S.K., Harper, D.A.T. & Portell, R.W. (2015) In deep water: a crinoid-brachiopod association in the Late Oligocene of Antigua, West Indies. *Lethaia*, **48**: 291–298.

Donovan, S.K., Harper, D.A.T. & Portell, R.W. (2017a) Shell-filled burrows in the Upper Oligocene Antigua Formation, Antigua, Lesser Antilles. *Ichnos*, **24**: 72–77.

Donovan, S.K., Harper, D.A.T., Portell, R.W. & Renema, W. (2014b) Neoichnology and implications for stratigraphy of reworked Upper Oligocene oysters, Antigua, West Indies. *Proceedings of the Geologists' Association*, **125**: 99 106.

Donovan, S.K., Harper, D.A.T., Portell, R.W. & Toomey, J.K. (2017b) Echinoids as hard substrates: varied examples from the Oligocene of Antigua, Lesser Antilles. *Proceedings of the Geologists' Association*, **128**: 326–331.

Donovan, S.K., Jackson, T.A., Dixon, H.L. & Doyle, E.N. (1995) *Eastern and Central Jamaica. Geologists' Association Guides*, **53**: i+62 pp.

Donovan, S.K., Jackson, T.A., Harper, D.A.T., Portell, R.W. & Renema, W. (2014a) Classic localities explained 16: The Upper Oligocene of Antigua: the volcanic to limestone transition in a limestone Caribbee. *Geology Today*, **30**: 151–158.

Felix, J. (1883) Die fossilen Hölzer Westindiens. *Sammlung palaeontologischer Abhandlungen*, **1** (1): 1–29.

CHAPTER 53

FIELD TRIP: *AKTUO-PALÄONTOLOGIE* OF SOUTHPORT BEACH, MERSEYSIDE

Preamble

This chapter is a bonus, for me as well as you. At the start of this book (p. xviii) I promised 52 chapters. Only at the proof stage did it become apparent that there would be a number of blank pages. My publisher, Anthony Kinahan, and I thought this would be unacceptable and plotted to fill the gap. The result is just one more field guide, but one a little different again; there are no rocks. Instead, it is an exercise in *Aktuo-Paläontologie*, that is, to define it imperfectly, the study of modern organisms and organic remains as if they were fossils, providing observations worthy of application to the fossil record.

> An aktuopalaeontologist is naturally interested in almost every aspect of marine biology, so many of which have their uses in palaeontology. However, he has certain additional interests that are normally not shared by his biological colleagues. (Schäfer, 1972, p. 2)

So, the intention is to examine modern remains (and extant examples of the same taxa, if available) as a key to the past, an endeavour that is truly Lyellian. Many of the studies of *Aktuo-Paläontologie* relate to taphonomy, palaeoecology or both.

Where might we do this? I mainly collect marine invertebrates, so a shelly beach would be ideal. The best place that I know in the British Isles for collecting modern shells that ask interesting taphonomic questions is Southport on the Irish Sea coast. At low tide in Southport, the sea retreats many hundreds of metres (Fig. 53.1), exposing sand and mud flats rich with shells. Every time that I go there I see something different; often it is something unusual and worth collecting.

How to get there

Southport is a seaside town and there is ample car parking for the motorist. As usual, I catch a train. There are direct services from Manchester and Liverpool. I take the train

Figure 53.1 A pier with a view. The beach at Southport near low tide, a magnificent expanse of sand and mud that is the graveyard of innumerable marine gastropods and bivalves, which lived offshore and have been transported eastwards after death.

at Salford Crescent and am in Southport in little over an hour. From Southport station to the beach is a walk of about 20 minutes west – the roads are such that it is almost a straight line. Wear your most waterproof boots as there will be some wading even at low tide. It always seems to be windy, so wrap up well.

Locality details

The beach at Southport, Merseyside, is on the Irish Sea coast of north-west England and below high tide. My shell collecting is commonly close to and seaward of the pier [NGR SD 328 180 and surrounding area] (Fig. 53.1). At the earliest, aim to arrive two to three hours after high tide; much of the beach will already be exposed. Southport has a particularly broad, sandy to muddy beach with no rock exposures, natural or otherwise, but some salt marsh (Gresswell, 1937). Dead, allochthonous valves of diverse bivalves and gastropods are common and varied, concentrated along the strand line and on the wave-rippled sand flats. Molluscs along the strandline will likely be ground down unless you time your visit to be just after a storm or particularly high tide, when complete specimens may be plentiful. The largest and most prominent gastropods are valves of the large whelk *Buccinum undatum* (Linné) and globular naticids, which may be clean, but are more commonly broken, bored, densely encrusted by balanid barnacles or serpulids, or a combination of these. Many of the dead whelk shells provide hard substrates that elevate the barnacles above the beach surface and form small 'balanid reefs' when the tide is in. The commonest bivalves are razor shells, *Ensis* sp. or spp.

What to look for

There are many observations to be made on Southport beach and collection is easy – take a big bag for specimens – but neither of these is worthwhile unless you are willing to interpret the remains. A good book for mollusc identification is a must (Beedham, 1972; Tebble, 1976). For example, disarticulated valves of *Barnea candida* (Linné) are not uncommon. This is a boring bivalve, but boring into what? Offshore, there must be a substrate suitable for this mollusc, perhaps a well-compacted mudrock or peat. Again, valves of the trough shell *Mactra corallina* (Linné) are commonly perforated by a single, conical boring. This is a predatory borehole, *Oichnus paraboloides* Bromley, and is likely the spoor of one of the common naticid gastropods found on the beach.

Let's move from generalities to specifics and examine the information provided by just two shells. All specimens illustrated in this chapter will be deposited in the Manchester Museum.

A bored whelk (adapted from Donovan, in press a)

Consider the figured whelk, *B. undatum* (Figs 53.2–53.6). This shell is an ichnological wonderland, providing evidence for organism–organism interactions of the sort that we might recognize in the fossil record. The substrate, *B. undatum*, is a common large benthic predator and scavenger in the subtidal environment around the British Isles. It is a species with a fossil record, such as in the Pliocene Red Crag Formation of East Anglia.

First, note the broken aperture (Fig. 53.2). Whelks with more or less broken apertures are a common feature of the beach at Southport. This specimen has a breakage that is both large and U-shaped. Such breakage may simply be mechanical damage to an allochthonous shell (Schäfer, 1972, fig. 239); it could be the spoor of a predatory crab (Schäfer, 1972, fig. 237); or it represents modifications made by a hermit crab that lived in the shell. There is no definitive evidence to indicate which of these is true in this example, but mechanical damage is always most likely. Whelk shells commonly show more or less damage to the outer (free) lip of the aperture. The present example is slightly unusual and may be due to the attentions of a crab. The shape of the breakage is suggestive of some predatory breaking and peeling back of the shell in a narrow band (Fig. 53.2); the width of the slot may indicate the width of the predator's claw. But the evidence is inconclusive, at best.

Entobia isp. is common at Southport; any robust shell is a good substrate for these branching borings of clionaid sponges. Small, round apertures are apparent above tiny chambers that are linked by slender canals (Fig. 53.3). *Entobia* canals and chambers have weakened the shell which has started to break through, as in Figure 53.3. Another good indicator of *Entobia* is a row of small apertures, as in the penultimate whorl seen in Figure 53.2.

Caulostrepsis taeniola Clarke is a U-shaped, straight or sinuous boring with a figure-of-eight aperture (Fig. 53.4). These are the borings of spionid worms, most likely *Polydora* sp.

Oichnus paraboloides Bromley is common on Southport beach in the form of predatory borings (see above), likely produced by predatory naticid gastropods (*Natica*

[Left] Figure 53.2 *Buccinum undatum* (Linné), lateral view of shell showing the broken outer lip of the aperture (after Donovan, in press a, fig. 1). Note also the healed breakage. Borings include *Entobia* isp. (small round holes) and figure-of-eight apertures of *Caulostrepsis taeniola* Clarke. Small breakages in the shell are probably the result of weakening by borings. The same specimen is figured in Figures 53.2–53.6. Scale in mm and cm.

[Right] Figure 53.3 *Buccinum undatum* (Linné), reverse side from Figure 53.2 (after Donovan, in press a, fig. 2). Borings mainly *Entobia* isp. The inverted Y-shaped structure marks the lines of canals broken through at the surface. Apertures are small and round, some broken through to reveal chambers. A triangular *Centrichnus eccentricus* is apparent slightly right of the centre line on the penultimate whorl. Scale in mm and cm.

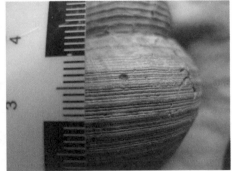

[Left] Figure 53.4 *Buccinum undatum* (Linné), figure-of-eight slot-shaped apertures of *Caulostrepsis taeniola* Clarke (after Donovan, in press, a, fig. 3). Scale in mm and cm.

[Right] Figure 53.5 *Buccinum undatum* (Linné), non-penetrative *Oichnus paraboloides* Bromley, circular and dish-shaped (after Donovan, in press a, fig. 4). Scale in mm and cm.

Figure 53.6 *Buccinum undatum* (Linné), triangular *Centrichnus eccentricus* Bromley & Martinell (after Donovan, in press a, fig. 5). Note the specimen is both depressed and has grooves concave towards the more pointed end, which is also perforated by *Entobia* isp. Scale in mm and cm.

spp.). But the boring in the whelk shell (Fig. 53.5) is non-penetrative and bowl-shaped. It may be an example of failed predation or, perhaps, *O. paraboloides* on this whelk shell may have been the domicile of some sessile, pit-forming invertebrate.

The most unexpected pits seen in this whelk are the rare triangular etchings known as *Centrichnus eccentricus* Bromley & Martinell (Fig. 53.6). It is a surface etching that is commonly attributed to the attachment scars of anomiid bivalves. In short, this whelk shell is rich in modern traces, some common (*Entobia*, *Caulostrepsis*) and others rare (non-penetrative *Oichnus*, *Centrichnus*). The pattern of breakage of the aperture poses the most interesting question of all – it is likely to be merely the mark of mechanical damage during transport on shore – or is it?

Barnacles and razor shells (adapted from Donovan, in press b)

But it isn't all modern traces. The specimen in Figure 53.7 is an articulated shell of the razor shell *Ensis siliqua* (Linné) with an intact ligament. The shell is about 170 mm long by 24 mm maximum width. At the posterior (siphonal) end the shell is encrusted on all external and internal surfaces by balanid barnacles, *Balanus crenatus* Brugiére. These are larger on the external surfaces than on the internal. Balanids are close packed and extend less than 25 mm from the posterior extremity. No evidence of encrustation is apparent on any other part of the shell. Such encrusted razor shells are rare, but I have collected six similar specimens from Southport so far.

This specimen is relevant to the study of the taphonomy of infaunal bivalve molluscs. The specimen (Fig. 53.7) has a pattern of infestation by balanids that is identical to a specimen of a different and alien invasive species, *Ensis americanus* (Binney), from the North Sea (Donovan, 2011, fig. 2). Razor shells are deep burrowers; the shell was undoubtedly encrusted post-mortem. The live shell moved to the surface shortly

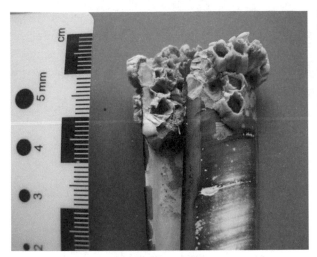

Figure 53.7 Left and right valves of articulated shell of *Ensis siliqua* (Linné) encrusted posteriorly on all external and internal surfaces by *Balanus crenatus* Brugiére (after Donovan, in press b, fig. 1). Scale in mm and cm.

before death in response to deteriorating environmental conditions, or, after death, the posterior was partly exposed by erosion of the surrounding sediment, or a combination of both. After loss of the soft tissues, but while the ligament was intact, the valves were constrained from gaping by the enclosing sediment.

The balanids may represent one spatfall despite those inside the shell being smaller than those externally. Balanids inside the razor shell were constrained by growing in an enclosed space. Further erosion exposed the shell completely, and it was carried onshore by tidal action. Throughout, the ligament remained unbroken.

To a palaeontologist, a valve encrusted both inside and outside by some shelly cementing organism would be interpreted as having had a long residence time on the seafloor. Encrustation on the inside of a valve would be interpreted as having probably occurred after rupture of the ligament. But this specimen demonstrates the disjunction between loss of soft tissues (days?) and loss of the ligament (weeks, perhaps months). Rather than being near-simultaneous, the ligament can persist for some time after the soft tissues are lost.

Balanids encrusting both the inside and outside of articulated shells have now been demonstrated in two species of razor shell, *Ensis* spp., and the cockle *Cerastoderma edule* (Linné) (Donovan, 2007, 2011; Donovan *et al.*, 2014, 2020). Although both are infaunal burrowers, razor shells and cockles are dissimilar in form and habit (Schäfer, 1972). If these disparate taxa can be encrusted both internally and externally before their ligaments rot, then it is reasonable to speculate that other infaunal bivalves may have been similarly infested both in the geological past and the present. Thus, a fossil disarticulated valve of a bivalve, encrusted internally by balanids or other shelly encrusters, did not necessarily reside on the sea floor for an extended period after death of the mollusc. The residence time might only have been weeks and it may have pre-dated the deterioration of the ligament.

The bivalve ligament soon dries and breaks if removed from the sea, but can persist in the marine environment (Schäfer, 1972, p. 164). No more than a rule of thumb exists, that a strong ligament may resist disarticulation for longer than a weak one (Brenchley & Harper, 1998, p. 87). It is therefore reasonable to suggest that *Ensis* has strong ligaments but thin, weak shells.

References

Beedham, G.E. (1972) *Identification of the British Mollusca*. Hulton Educational Publications, Amersham.

Brenchley, P.J. & Harper, D.A.T. (1998) *Palaeoecology: Ecosystems, Environments and Evolution*. Chapman & Hall, London.

Donovan, S.K. (2007) A cautionary tale: razor shells, acorn barnacles and palaeoecology. *Palaeontology*, **50**: 1479–1484.

Donovan, S.K. (2011) Post-mortem encrustation of the alien bivalve *Ensis americanus* (Binney) by the barnacle *Balanus crenatus* Brugière in the North Sea. *Palaios*, **26**: 665–668.

Donovan, S.K. (in press a) Fossils explained: An exciting, but bored whelk. *Geology Today*.

Donovan, S.K. (in press b) Fast post-mortem encrustation of razor shells: examples from the Irish Sea and palaeontological implications. *Proceedings of the Yorkshire Geological Society*.

Donovan, S.K., Cotton, L., Ende, C. van den, Scognamiglio, G. and Zittersteijn, M. (2014) Taphonomic significance of a dense infestation of *Ensis americanus* (Binney) by *Balanus crenatus* Brugière, North Sea. *Palaios*, **28** (for 2013): 837–838.

Donovan, S.K., Hoeksema, B.W., Fransen, C.H.J.M., Vonk, R. and Adema, J.P.H.M. (2020) Unusual preservation of North Sea shells: Scheveningen, North Sea coast, the Netherlands. *Bulletin of the Geological Society of Norfolk*, **70**: 55–65.

Gresswell, R.K. (1937) The geomorphology of the south-west Lancashire coast-line. *Geographical Journal*, **90**: 335–349.

Schäfer, W. (1972) *Ecology and Palaeoecology of Marine Environments*. University of Chicago Press, Chicago.

Tebble, N. (1976) *British Bivalve Seashells*. Second edition. HMSO, Edinburgh.

Glossary

A

acorn barnacle: Conical to cylindrical, multi-plated, gregarious encrusting organisms. They are arthropods, that is, related to trilobites, crabs and insects. The unmineralized arthropod body of the barnacle was inside the conical shell and is commonly not preserved, but the shells are common fossils in Cenozoic deposits (Fig. 3.1).

allochthonous: Fossil organisms or other clasts that have been transported. Such transport is commonly laterally, with shells being displaced by bottom currents or carcasses sinking to the bottom post-mortem, or vertically, with shells being displaced by burrowing organisms. See also *autochthonous* and *parautochthonous*.

ammonite: An extinct group of nektic marine molluscs related to extant octopus and squid (the cephalopods). Ammonites ranged from the Late Palaeozoic until the end of the Cretaceous; there is now strong evidence that some survived into the earliest Cenozoic. The ammonite shell was chambered and is commonly (but not invariably) coiled. They are distinguished from the closely similar nautiloids, which are still extant, by having complexly folded suture lines where the walls of the chambers met the inside of the shell. Suture lines of nautiloids are simple.

ammonium chloride puffer: A device for coating a fossil with a thin, matt white layer of ammonium chloride (Fig. 28.2). As explained in Chapter 28, this has the advantage of giving the specimen a uniform look, enhancing its appearance under the camera.

angular unconformity: A major break in a rock succession, commonly representing millions of years. In a sedimentary rock sequence, the plane of an angular unconformity is a major discordance with the beds below angled to those above (Fig. 6.2).

annelid: Members of the Phylum Annelida are the segmented worms, such as the terrestrial earthworms in your garden. In the marine fossil record annelids are most commonly encountered as tubes, commonly secreted by some groups that encrusted shells.

anoxic: Without oxygen.

ash fall: Volcanoes erupt gases, liquids (lava) and solids (ash). The ash may become distributed over a broad area as a sedimentary bed by the wind. It may also become concentrated on the shallow shelf areas around a land mass, and then being moved as one flow into deep water (see turbidity current) by, for example, a major storm or earthquake.

autochthonous: Organisms preserved where they existed in life. This is not the same as a life assemblage, which preserves organisms that lived together, although not necessarily in that place. For example, consider a Cretaceous nektic life assemblage – ammonites, fishes, mosasaurs – which may be preserved together on a bed even though they lived elsewhere, in the water column. See also *allochthonous* and *parautochthonous*.

autotomy: Self-mutilation. Most notably, this is practised by the echinoderms such as crinoids and echinoids. For example, if a predator attacks an arm, a crinoid may autotomize this organ and, over time, grow a replacement.

B

balanid, see **acorn barnacle**.

bed: Sedimentary rocks are organized into beds, at least on the scale of the exposure. A bed is bounded by bedding planes, one each at the top and bottom. It is the beds we see, but it is the bedding planes that probably represent a greater interval of time. Beds are where we find fossils.

bedding plane, see *bed*.

benthic: A benthic marine organism lives on the sea floor. It may be sessile (stationary) like a colonial coral or vagile (moves around) such as a crab.

biostratigraphy: The correlation and sequencing of the succession of sedimentary rocks according to their fossil content. The fossil record is arranged in a set order, so that a bed with a woolly mammoth will always be higher in the sedimentary succession ('younger') than beds bearing dinosaurs.

biostratinomy: The study of that part of a fossil's history between death (including cause of death) and final burial. Essentially, this is an examination of the carcass (mainly the skeleton) as a sedimentary particle.

blastoid: An extinct group of stalked echinoderms.

boring: A hole made into or through a hard substrate, such as shell, rock, bone or wood. The mechanism of a borer may be chemical (dissolution), physical (with a hard grinding organ) or a combination of both.

brachiopod: A group of bivalved invertebrates (that is, the shell is formed from two pieces or valves) that were common in the Palaeozoic (Figs 24.2, 45.3 and 52.6), but which have been a minor component of the benthos following the end Permian mass extinction (Fig. 52.6). They are distinct in their geometry from infaunal bivalve molluscs. In brachiopods, the valves are unequal (inequivalve) and bilaterally symmetrical; in burrowing bivalves, the valves are equal (indeed, mirror images, more or less) and asymmetrical.

breccia: A sedimentary rock composed of angular clasts greater than 4 mm in diameter (Table 7.1). The clasts are commonly of varying composition (= polymict). See Chapter 7.

burial: A specimen becomes buried during death (buried alive) or subsequently. While buried, the surrounding rock will be lithified, and the chemistry of the rock and specimen may be altered chemically (diagenesis).

C

camera lucida: A drawing tube associated with a binocular microscope, enabling you to see simultaneously both the specimen of interest and your hand. With a pencil, it is possible to draw the fossil as if it were being traced from a photograph. See Chapter 29.

cementation: As well as being compacted (see below), sediment grains may be cemented together, forming a sedimentary rock. The cement is precipitated from percolating ground water and may be one or more of a diversity of minerals, commonly including quartz or calcite.

chert: A chemical sedimentary rocks that forms after deposition of a rock sequence, commonly in limestones. Flint is a chert found in Chalk. Chert is a precipitate from silica carried in solution. The origin of the dissolved silica is siliceous fossils dissolved in the alkali groundwaters pervasive in lime-rich successions, including sponges and certain microfossils. Cherts may enclose fossils that can be dissolved out subsequently (*q.v. Steinkern*, screwstone).

clast: Grains that make up siliciclastic sedimentary rocks. My own preference is to call fragments in sandstones and more fine-grained sedimentary rocks grains, and in more coarse-grained sedimentary rocks clasts, but they are, in truth, all clasts and all grains.

coal: This is a sedimentary rock formed from vegetable matter in anoxic conditions. A leaf will fall off a tree, turn brown, dry out and turn to dust; this happens in the presence of oxygen. If it falls into, for example, a bog with anoxic (*q.v.*) waters, it breaks down more slowly without losing its carbon as carbon dioxide; there is no oxygen. A thick deposit of such carbon-rich organic fossils is a coal.

colonial coral: Corals can be individual (= solitary) or colonial, formed by the splitting into two or more branches. Of the major fossil stony coral groups, tabulates (Palaeozoic) are all colonial (such as Fig. 10.1, 45.2); the rugose (Palaeozoic) and scleractinian (post-Palaeozoic) corals include both colonial and solitary species (Fig. 45.1).

compaction: The reduction in volume of a sediment as it is shaken (earthquakes) and squashed (weight of overlying beds, called overburden). This can be modelled with a jam jar of dry sand filled to the brim. Screw the lid down. Shake the jar as hard as possible and for as long as possible. Remove the lid and note that your model earthquake has reduced (= compacted) the volume of sediment.

concretion, see *nodule*.

conglomerate: A sedimentary rock composed of rounded clasts greater than 4 mm in diameter (Table 7.1). The clasts are commonly of varying composition (= polymict). Fossils in clasts are derived from an older rock; fossils in the sedimentary rock between clasts are likely to be contemporary with the deposition of the rock. See Chapter 7.

crinoid: My favourite group. Crinoids (sea lilies) are stalked echinoderms which are plant-like in appearance, but are animals analogous to a starfish with the mouth directed upwards (away from the seafloor) and commonly held above it by a multiplated, segmented column. They are rarely preserved complete, disarticulating into component plates soon after death. They are commonly preserved as separate plates of the column, called columnals, or lengths of column, called pluricolumnals. See also *autotomy*.

cross-bedding: A sandstone bed is commonly bounded by upper and lower bedding planes that are often parallel. Cross-beds at an angle to these surfaces are evidence for lateral movement of water-borne sand. These are common in many sedimentary successions, associated with energetic deposition. Analogous deposition in the terrestrial environment may produce sand dunes in a desert environment.

D

death assemblage: A group of fossils in a bed and/or on a bedding plane that did not live together in life and became mixed post-mortem. For example, to give a real example, the Miocene Grand Bay Formation of Carriacou in the Lesser Antilles preserves a range of organisms from the terrestrial environment (land snails) to open ocean plankton, washed together by turbidity currents (*q.v.*) in deep water.

diagenesis: The changes, mainly chemical, that occur to a fossil after final burial. These can vary from very little, such as when a calcite (calcium carbonate) shell is buried in a limestone, to dissolution, recrystallization and chemical replacement.

disarticulation: The falling apart of multi-element skeletons, commonly post-mortem, but it can happen before death, such as leaves falling from a tree. Common organisms with complex multi-element skeletons that have a good fossil record include plants, vertebrates, echinoderms and arthropods.

E

echinoderm: The spiny-skinned animals – crinoids, echinoids, starfishes, brittle stars and sea cucumbers. Only the first two groups are common fossils (*q.v.*). See also *autotomy*.

echinoid: A sea urchin, a member of the phylum Echinodermata with a globular, flattened or heart-shaped test (shell). First appeared in the Upper Ordovician and diversified in the post-Palaeozoic. See also *autotomy*.

encruster: An epifaunal (*q.v.*) organism that cements to a hard substrate (Figs 3.1, 3.2).

epifaunal: A general term describing those organisms that live or lived on the surface of the sediment or on the body of another organism. The former would include most brachiopods and all crinoids; the latter might include acorn barnacles (*q.v.*) or a parasitic snail (Fig. 22.2E).

exposure: The part of a rock's outcrop (*q.v.*) that is seen at the surface. Fossils are collected from exposures – quarries, road cuts, cliffs at the coast, man-made trenches and so on.

ex situ, see *float*.

extrusive: Lavas, gases and volcanic ashes are extrusive products of the Earth's inner heat, all erupted from volcanoes. An extrusive deposit may fossilize organisms, for example, by burial in an ash fall.

F

flint: A secondary, siliceous sedimentary rock found in Chalk. A chert (*q.v.*).

float: Specimens found *ex situ*, that is, removed from their parent bed and found elsewhere. For example, consider a quarry with a scree (*q.v.*) slope developed against the face by loose rocks relocated downslope by gravity. Fossils in this scree pile will be float specimens, bed unknown. Determination of exactly which bed it came from may be possible by careful investigation.

foraminifera: Single-celled microfossils – protists – with a hard skeleton, commonly either calcite or agglutinated grains. Either planktonic or benthic; benthic forams (to use their common abbreviation) may be large enough to see with the naked eye. The largest that I have ever seen, on display at the Geological Museum of the University of the West Indies, was as wide as a saucer, but thin. Forams are important biostratigraphic markers.

fossilization: The conversion of a dead organism into a fossil. This may involve multiple pathways. For example, a fossil has typically lost its unmineralized ('soft') tissues, but a woolly mammoth may be preserved with its tissues and even fur intact in Siberian permafrost, and is certainly an exceptional fossil. Most body fossils preserve hard skeletons only; if soft tissues are preserved they will likely be mineralized, rapidly, during diagenesis. Typically, but not invariably, fossilization will involve some chemical changes in the hard skeleton.

functional morphology: The study of how a fossil worked as an organism.

G

geopetal infill: Called a 'palaeo-spirit level' by some as an indicator of the horizontal at the time of deposition. A typical example might be an empty shell into which some sediment leaked and became lithified. The flat top level of this sediment surface defines the horizontal at the time of deposition (Fig. 10.4).

H

heart urchin: A heart-shaped echinoid (sea urchin) (Figs 2.1, 24.1). A typical example is the well-known Chalk *Micraster* L. Agassiz (Fig. 28.1D–F).

holotype: Description of a new species of fossil requires definition on the basis of a defined suite of one or more than one specimens, called the type series. One of these, regarded as the most typical of the group, is called the holotype; if there are two or more specimens, the others are paratypes (see Davies, 1972, chapter 13). These specimens are registered in a museum or university collection so that they are available for consultation for other palaeontologists, both now and in the future.

I

induration: The hardening of sedimentary rocks (*q.v.* lithification), which may be due to one or a combination of factors, such as cementation, compaction and/or heat, due to depth of burial or the proximity of an igneous body.

infaunal: A general term describing those organisms that live or lived beneath the sediment surface such as, for example, burrowing bivalve molluscs.

in situ: In its original place, that is, autochthonous.

intrusion: An igneous body that has solidified in a sequence of crustal rocks as it cooled down. Intrusions take many forms, commonly cross-cutting bedded rocks except where they force themselves between beds when still hot and liquid, and subsequently solidify (= sill).

L

Law (or Principle) of Superposition: In any sequence of layered rocks (sedimentary beds, lava flows), unless there is reason to suspect that it has been inverted by tectonic action (Earth movements during mountain building), the oldest bed will be at the bottom and the youngest bed will be at the top. Thus, examining beds from bottom to top will be to scrutinize a progression analogous to the time taken for them to be formed. This is also known as Steno's Law.

life assemblage: A group of fossil organisms, preserved in close association, that reflects an original relationship that they had when alive.

limestone: A rock rich in lime (calcium carbonate, $CaCO_3$), commonly produced in marine environments, particularly those rich in shelly organisms, but which can also be formed in freshwater, such as lakes, and terrestrial settings.

lithification: The diverse processes whereby a sediment is turned into a sedimentary rock (*q.v.* induration). These include compaction (from the weight of the accumulated overlying sediment pile) and cementation (both from minerals deposited from percolating groundwaters, and chemical changes of the included mineral and rock grains).

M

mafic: An igneous rock formed by predominantly dark-coloured silicate minerals, rich in iron and magnesium.

metamorphic: A metamorphic rock – originally a sedimentary, igneous or metamorphic rock – is one that has had its mineralogical and physical attributes altered by extreme heat and/or pressure within the Earth, commonly associated with mountain building (associated with deep burial; heat and pressure) and igneous activity (heat only).

microfossil, see *micropalaeontology*.

micropalaeontology: The study of microfossils. Typical microfossil groups include foraminifera and ostracods. Macroscopic members of these groups are still studied by micropalaeontologists despite their size.

monospecific: Comprised of a single species. This might refer to a genus including just one species or a naturally occurring assemblage in which all specimens belong to a single species.

mudrock: A rock composed of mud-sized grains (less than 1/256 mm in size; Table 7.1), such as mudstone and shale.

mudstone: A massive, non-fissile mudrock.

N

nektonic: Free-swimming aquatic organisms, such as fishes and marine reptiles.

New Red Sandstone: Widespread red beds deposited in the Permian to Triassic (= Permo-Triassic; Table 6.1) interval in the British Isles and elsewhere. Apart from Permian marine deposits in north-east England and contiguous areas, all deposits in Britain of this interval are very poorly fossiliferous red beds, except locally.

nodule: A nodule or concretion is a structure precipitated after formation of a sedimentary rock. The nodule commonly has a composition in contrast to the host rock, such as calcite, silica (chert) or phosphate. Growth of a nodule is due to a chemical contrast between the host rock and its percolating ground water.

O

Old Red Sandstone: Red beds – that is, terrestrial deposits – of Devonian age (Table 6.1), widespread in many parts of the British Isles apart from in south-west England. Old Red Sandstone is an older term than Devonian. The type area of the Devonian is a mainly marine succession. Poorly fossiliferous except locally.

outcrop: The total surface expression of a rock formation or other rock unit. By surface, it is meant that this would be as seen if all cover – soil, buildings, vegetation, etc. – were removed. Most exposures are much smaller than the outcrop.

P

palaeoautecology: The study of the palaeoecology of individual fossil organisms, particularly: their functional morphology and; its relation to their environment.

palaeodepth: The water depth at which an ancient aquatic deposit was laid down. Such determinations are at least partly qualitative, the principal lines of evidence including sedimentary rock type and structure, and body and trace fossils.

palaeoecology: The ecology of fossil organisms, including their own habit and habitat, and the relationship to all aspects of the biological, chemical and physical environment.

palaeosynecology: The study of communities (two or more individuals) of the past, and their relationships to each other and their environment.

palaeotemperature: The study of temperature as a component of an ancient physical environment. It can be determined chemically, by analysis of stable oxygen isotopes, or qualitatively from inferences deduced from palaeontology and sedimentology. See Chapter 20.

paratype: If a new species is described on the basis of two or more specimens, only one can be designated as the holotype (*q.v.*). All other specimens upon which the original description is based are also part of the type series and are titled as paratypes.

parautochthonous: Organisms preserved close to where they existed in life. Final burial likely followed death closely and after only minimal transport. For example, there is a common suspicion, which can often be demonstrated, that shells preserved on a shallow continental slope (such as eastern North America) will have undergone little transport, at most, before final burial. See also *allochthonous* and *autochthonous*.

Permo-Triassic: The Permian plus Triassic, largely represented in the British Isles by the New Red Sandstone (*q.v.*). Because the New Red Sandstone is largely unfossiliferous, it cannot be subdivided on the basis of biostratigraphy and is commonly 'lumped' into Permo-Triassic until, and if, more definite correlative data becomes available.

Phanerozoic: Time since the start of the Cambrian and all the rock succession deposited subsequently. It is this part of the geological record that records the presence of plentiful life, and includes the Palaeozoic, Mesozoic and Cenozoic.

photographic stand: A device for securing a camera in a steady mount of known orientation. Such a stand is an important adjunct to the photography of fossils (Fig. 28.3).

planktonic: Free-floating organisms in an aquatic ecosystem. Many planktonic organisms are microfossils, which can be small and light, such as foraminifers. Ancient macroscopic plankton commonly maintained a flotation device, such as in graptolites (Prothero, 2020, fig. 18.3).

pluricolumnal, see *crinoid*.

polyspecific: Comprised of more than one species. This might refer to a genus including two or more (sometimes many more) species or a naturally occurring assemblage in which specimens belong to a range of taxa.

preservation: Of the many organisms that have lived on the Earth, only a sample is found in the fossil record. Preservation of a once-living organism in the rock record follows myriad chance pathways of biological, chemical and physical modification. This results in a fossil, derived from a dead organism and sufficiently alike its original form to be worthy of study, but nonetheless altered post-mortem.

puffer, ammonium chloride, see *ammonium chloride puffer*.

R

recumbent: Lying down. Some groups of organisms, commonly erect, include members that lived recumbent on the sediment surface, such as certain members of the crinoids and solitary corals.

red beds: Terrestrial sedimentary deposits, commonly siliciclastics, stained red by the presence of common iron in the ferric (iron III) state. The most notable red beds in the British Isles are the Old Red Sandstone (Devonian) and New Red Sandstone (Permo-Triassic) (*q.v.*). Neither is commonly fossiliferous.

reef: A word with a plethora of definitions. For our purposes, a reef is an organic structure growing up from the seafloor, forming an ecological association of diverse sessile and vagile benthic organisms (commonly fossilized) with nektonic groups such as cephalopods and fishes.

residence: The residence time of a carcass on the seafloor can vary from zero seconds (buried alive) to years. A bone or shell on the seafloor offers many possibilities for other organisms to dwell on (encruster) or in it (borer). Given sufficient time, a succession of encrusting shelly forms may develop, younger shells overgrowing those that have died.

reworked fossil: Fossils that are eroded out of their 'native' beds and redeposited in a younger formation are called 'reworked'. For example, Late Cretaceous rudist bivalves (*q.v.*) reworked from shallow water limestones are known from deeper water siliciclastics in the Paleogene of Jamaica.

rudist: The rudists were a group of aberrant bivalve molluscs with robust shells that range from the Jurassic to mid-Cretaceous. They are amongst the largest bivalves known, but most were strongly inequivalve. They were gregarious and formed shell banks analogous to those of oysters, but were not reef builders as has been reported in the literature for many years.

S

sandbox: A homemade device, literally a box of dry sand, used to arrange a fossil in any orientation while, say, the glue uniting two broken parts dries.

sandstone: A siliciclastic rock, grain size between 1/16 and 2 mm (Table 7.1), and with grains commonly composed of quartz, at least in part.

scree: A superficial deposit of loose clasts of assorted sizes jumbled together, and formed due to the physical erosion of a quarry face or cliff. The scree is commonly banked up against the face, so at least a coarse determination can be made of where any given clast may have originated.

screwstone: A chert (*q.v.*) in which crinoid columnals and pluricolumnals are preserved as moulds. Long pluricolumnals with the central axial canal infilled superficially resemble long screws (Donovan *et al.*, 2016, fig. 2c).

sedimentary: Pertaining to sediments, the products of physical and chemical breakdown or chemical precipitation (including organic action) formed at the Earth's surface (Chapters 7, 8). Through the action of burial (heat, pressure) and chemical precipitation (cement), sediment is converted into a sedimentary rock.

sedimentary bed, see *bed*.

sessile, see *benthic*.

sexual dimorphism: In some species, the two sexes are different in form. We know this as it applies to humans, but it is also found in some fossil invertebrates such as ammonites and ostracods.

shale: A fissile mudrock (*q.v.*). The fissile property of a shale is due to bedding; slates (*q.v.*) are metamorphosed mudrocks (mudstones or shales) that develop cleavage, a physical feature that is not analogous to, or commonly parallel with, bedding.

siliciclastic rocks: Sedimentary rocks composed of grains derived from pre-existing rocks (sedimentary and/or igneous and/or metamorphic) that were rich in silicate minerals, particularly quartz (Chapter 7).

silicification: A process whereby an original fossil of any composition is replaced by silica. Well-known examples include the flint *Steinkerns* (*q.v.*), which are siliceous internal moulds of Chalk echinoids preserved in flint.

siltstone: A siliciclastic sedimentary rocks with grains 1/256 to 1/16 mm in diameter. That is, a siltstone is coarser grained than a mudrock and finer grained than a sandstone.

slate: A low-grade metamorphic rock with a distinctive cleavage and formed from a sedimentary mudrock in mountain-building episodes. Robust fossils may still be preserved in a slate, albeit deformed.

spatfall: The mass settlement of larvae of marine benthic organisms, such as barnacles or oysters, resulting in a monospecific (*q.v.*) assemblage of closely associated shells of similar age.

spreite: U-shaped structures either within the 'arms' of a U-shaped burrow or beneath the curved base, indicating movement down or growth, or movement up, respectively (Fig. 24.3).

Steinkern: Siliceous internal moulds of flint, commonly in Chalk echinoids.

stratigraphy: The study of the identification, ordering and correlation of beds or strata.

T

taphonomy: The study of the preservation of fossils, including cause of death, post-mortem changes before burial and diagenetic changes following final burial.

theca (plural thecae): The cup of a crinoid, blastoid or cystoid. The theca may incorporate plates of the column or arms into its structure, best seen in many of the camerate crinoids (Fig. 50.4A).

trace fossil: The two common types of fossils are body fossils, such as parts of the hard skeleton such as a shell, tooth or bone, and trace fossils. The latter are evidence of organic activity, called by some 'frozen ecology', and include such artefacts as tracks and trackways (footprints), trails, burrows, borings and coprolites. It is rare that a trace fossil and its producing organism are preserved in close association. These can provide unique information without the producing organism being confidently identified.

transport: The movement of a fossil organism either after death or, in some instances, as the cause of death, such as being swept away and buried alive (see, for example, Fearnhead *et al.*, 2020).

trilobite: A group of extinct arthropods common in Palaeozoic rocks because the chitinous exoskeleton was partly calcified, favouring preservation.

tuffaceous: Tuff is a rock made from compacted volcanic ash.

turbidity current An underwater 'landslide' in which sediment is rapidly moved downslope by gravity and under its own weight. A trigger for a turbid flow might be an earthquake or major storm. Deposition in deep water follows particular patterns based on grain size and energy. See *ash fall*.

type: One or more specimens that are used to define the unique characters of a new species. See *holotype* and *paratype*.

V, W

vagile, see *benthic*.

volcanic: A product of a volcano, such as lava or ash.

way-up structure: A feature of a rock succession that defines the orientation at the time of deposition. These can be sedimentological (cross beds), palaeontological (fossils preserved in life orientation), volcanic or structural. See also *geopetal infill*.

References

Davies, A.M. (revised Stubblefield, J.). (1972) [tenth impression]. *An Introduction to Palaeontology*. Third edition [edition first published 1961]. Thomas Murby, London.

Donovan, S.K., Jagt, J.W.M. & Deckers, M.J.M. (2016) Reworked crinoidal cherts and screwstones (Mississippian, Tournaisian/Visean) in the bedload of the River Maas, south-east Netherlands. *Swiss Journal of Palaeontology*, **135**: 343–348.

Fearnhead, F.E., Donovan, S.K., Botting, J.P. & Muir, L.A. (2020). A Lower Silurian (Llandovery) diplobathrid crinoid (Camerata) from mid-Wales. *Geological Magazine*, **157**: 1176-1180.

Prothero, D.R. (2020) *Fantastic Fossils*. Columbia University Press, New York.

Index